Seeing Gent
A City Through Its Windows

Introduction

The International Masters at KU Leuven draws students from all over the world to study on its English language course. Part of the attraction is the opportunity to study in Gent, a beautiful historic city famed for its architecture. Much remains of the medieval city that is characterised by tightly packed window-wall façades. In devising the Building Technology Workshop, I was keen to focus attention on this extraordinary resource: to closely observe, record and represent how a traditional European city is made. The requirements to draw by hand and make physical models sought to exercise the connection between eye and hand which can be of such great use to architects. The variety of the workshop output is testament to Gent's heterogeneity and an equally delightful introduction to each student's way of seeing.

David Kohn
Visiting Professor 2014/15

Contents

Project	Page	Project	Page
Group 1	**6**	**Group 4**	**66**
Patershol	8	Vlaamse Opera	68
Huis van Alijn	10	Artevelde Hogeschool Kantienberg	70
Corner Beer Shop	12	Hotel Falligan / Literary circle	72
Private House	14	De Oudburger	74
Drongenhof Kapel	16	Stadhuis	76
Vooruit	18	Private House	78
Group 2	**20**	**Group 5**	**80**
Zebrastraat	22	Booktower	82
Private Building	24	Sint-Barbaracollege	84
Corner House G-S	26	Café het Spijker	86
Design museum Gent	28	Private House	88
Studenthome	30	Frank Steyaert's Atelier	90
Private House	32	Fallen Angles store	92
Florist's	34	Desmet-Guequier cotton mill	94
Twiggy store	36	Poezie centrum	96
Shelter 7	38	Nucleo Kunstenaars atelier	98
Xi / Café Costume	40	Coffee Max	100
Stam	42	Platte beurs coffee	102
Group 3	**44**	**Group 6**	**104**
Ramen	46	"Our House" Socialist Worker	106
Private House	48	Groot gerechtsgebouw	108
Bakerij Himschoot	50	Abby from Sint-Lucas	110
HoGent-Bijloke Campus	52	UFO Gent	112
Masons' Guild Hall	54	Les Ballets C de B	114
Gravensteen Castle	56	Private House	116
Pakhuis restaurant	58	Sint Pieter Station	118
The House of Alijn	60	Proud of House	120
Campus KTA Casinoplein	62	Old courthouse	122
Les Ballets C deB and LOD	64	Pakhuis Clemmen	124

Group 01

Patershol
Huis van Alijn
Corner Beer Shop
Private House
Drongenhof Kapel
Vooruit

BUILDING: Patershol
LOCATION: Kaatsspelplein 8 Gent
STUDENT: Barbera Poláková

Building was part of the historical complex with Drongenhof Kappel in Patershol. It's foundation dates to 17.century. The original purpose of the building isn´t known. Later, in years 1987-1988 was the object under massive reconstruction. From this period comes also a design of current windows. Building was renovated 2.nd time and newly opened on 2.10.2012. The Pratershol is now a meeting place for people over 55 years. It is owned by the city of Gent and is run by volunteers. It's windows in turqoise color and modernistic style are in interesting contrast with the historical background.

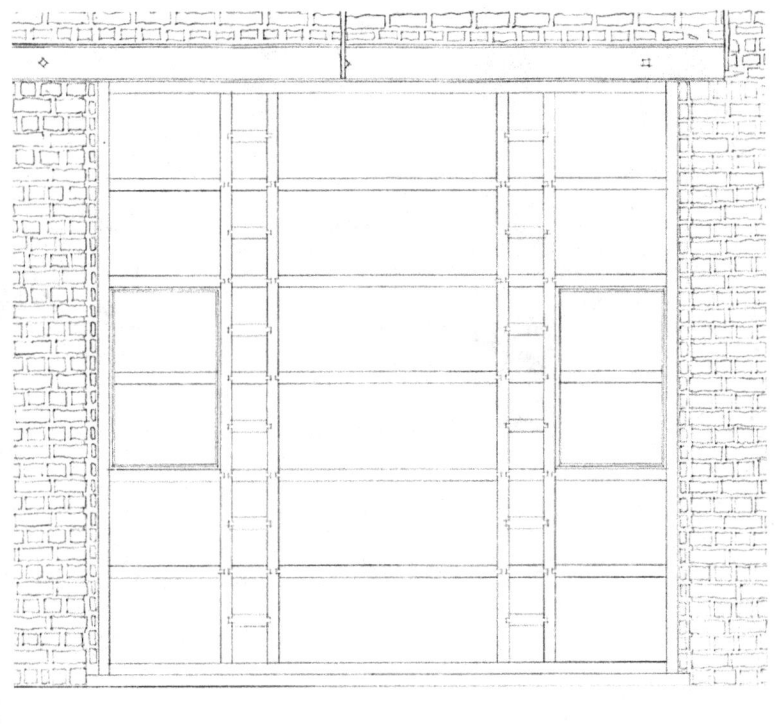
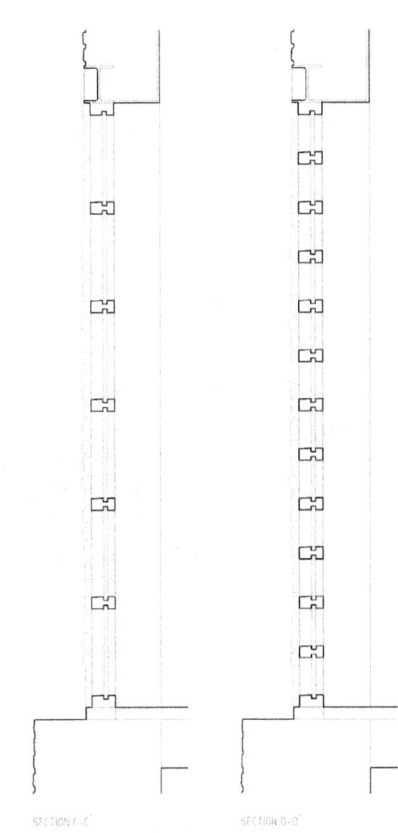

SECTION C-C'   SECTION D-D'

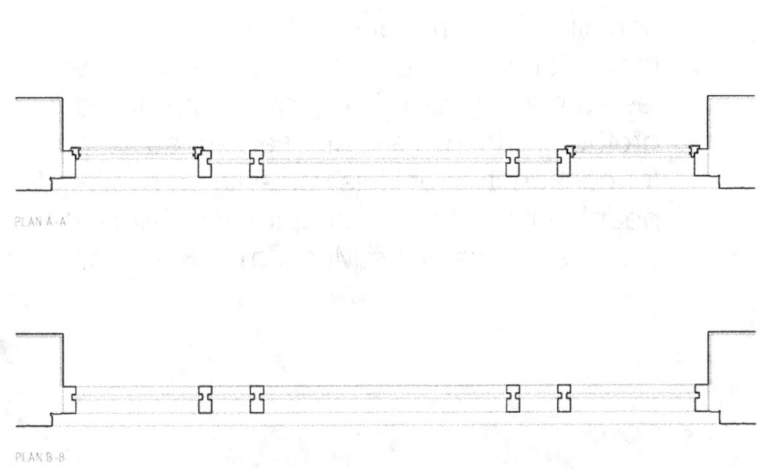

PLAN A-A'

PLAN B-B'

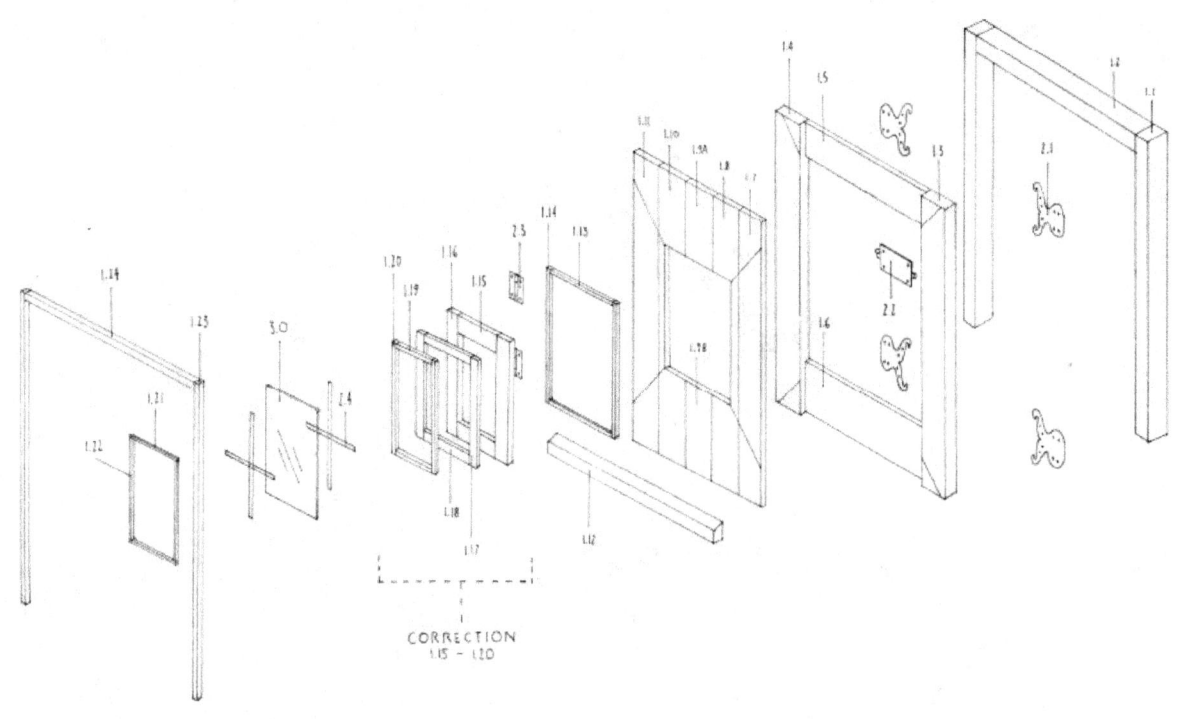

BUILDING: Huis van Alijn
LOCATION: Kraanlei 65
STUDENT: Iwo Borkowicz

The building was constructed in 1363 after reconciliation of the feud between two Gent aristocrat families, the Rims and the Alijn, which lasted many years. After having murdered Alijn brothers, the Rijms were sentenced to pay a for construction of Huis van Alijn - a sanitary for people in need and a hospital for children. Nowadays it houses the Museum of Folklore.

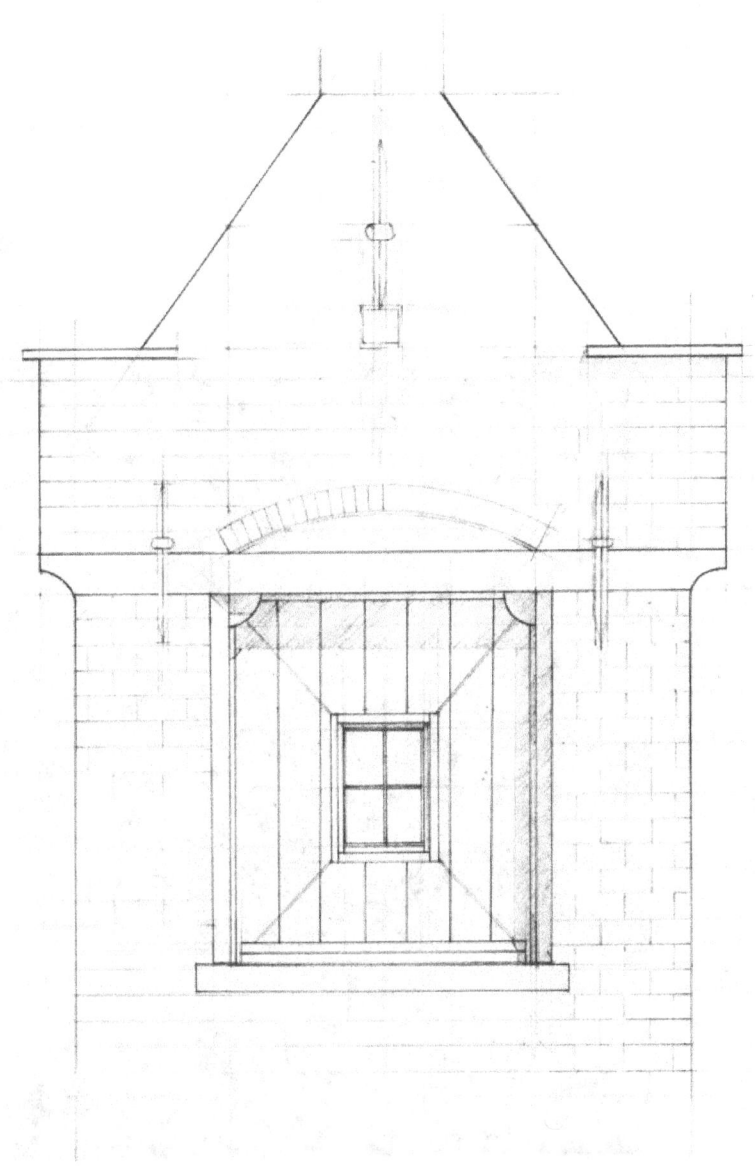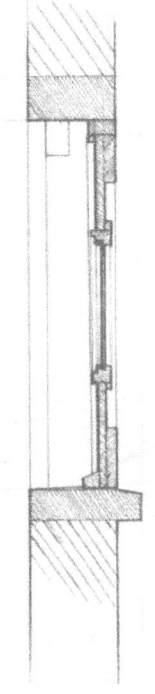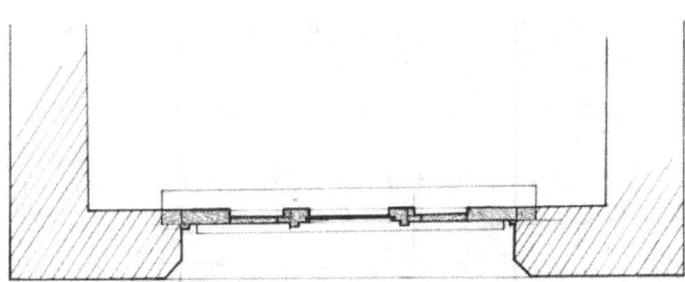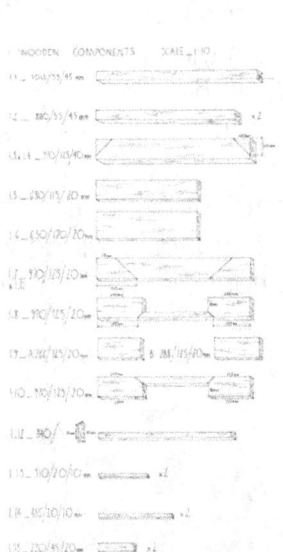

BUILDING: Corner Beer shop
LOCATION: Sint Veerleplein Gent
STUDENT: Gonçalo Alves Oliveira

The selected window is located in the heart of Gent, at Sint Veerle square.
This window works like a frame, showing all the beers like a painting; these beers tell a story of the country, representing all the cities and its traditions. Through it, from the inside of the store, we can look directly at the castle, symbol of Gent and an important part of the Flanders' history. The colour, the shape, the location and the content on the inside and out, were all principal ingredients to select this specific window, and to rethink it as an important piece of the history of Gent.

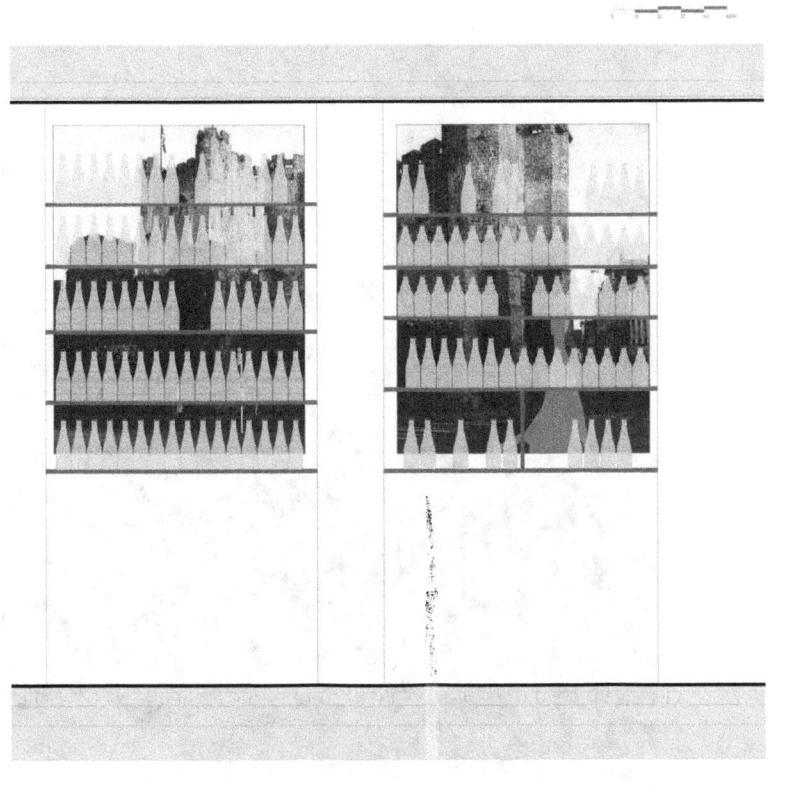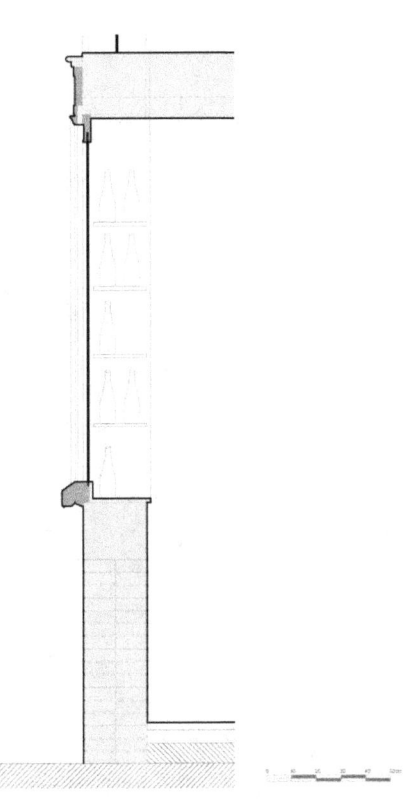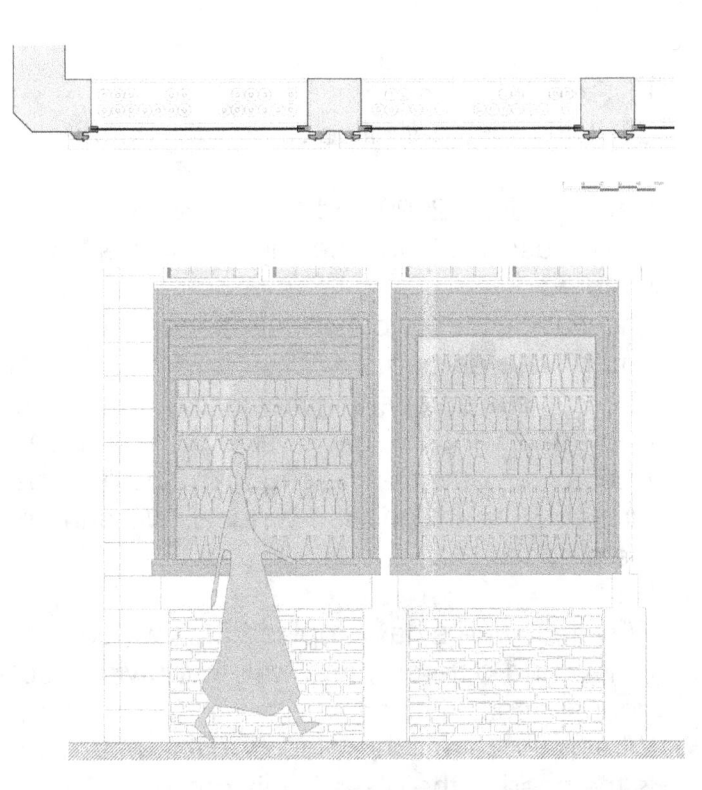

BUILDING: Private House
LOCATION: zilverhof
STUDENT: Hélder Filipe Gonçalves Ferreira

Between June 2009 and November 2014.
The house and an opening in the wall were restored...
The wall divides two outdoor spaces, the water and private garden.
The wall has an opening that in 2009 was a half degraded window that in time was replaced. Another window that belonged to another opening. It is an improvised window. We can see the patches.
But...
Why the owners have replaced the window (for a new and different) in a wall that divides outdoor spaces?
What is the role of this window in this wall? What is the need of the glass in this context?
Why preserve this opening with a window?

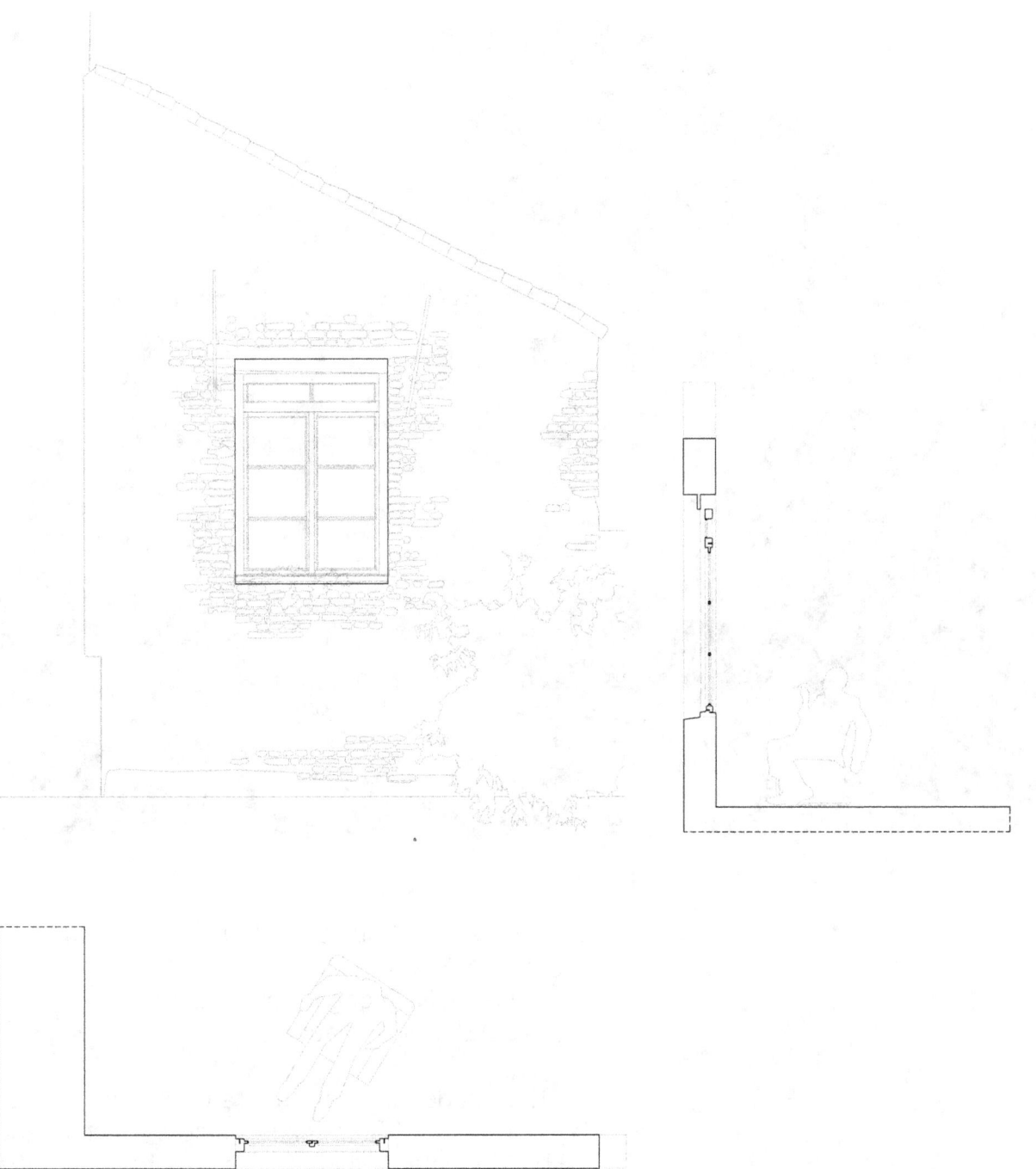

BUILDING: Drongenhof Kapel
REALISATION: 1602
LOCATION: Drongenhof 32 Gent
STUDENT: Igor Machata

The chapel of the former refuge of the Norbertine abbey of Drongen is a single-aisled chapel build in late Gothic style. It was build in 1607. This late Gothic chapel is a protected monument. It has been purchased by the city of Gent and is still awaiting restoration. For a long time it is only serving as a room for temporary exhibitions, warehouse, garage. Only fragments are left from the original accidence of the gothic windows. All of the windows have been walled. But the one that I chose is somehow different, more important. It has been closed, and then probably reopened again for some reason. It has a story in it.

SECTION B-B'

1 meter

BUILDING: Vooruit
ARCHITECT: Ferdinand Dierkens
REALISATION: 1911
LOCATION: Sint-PietersnieuwstraatGent
STUDENT: Maja Berden

The reason I was triggered to choose the window in the Vooruit Festivities Hall café is because of the interesting historical context of the building and today's relevance of the building. Vooruit Festivities Hall was built between 1910-1913 by architect Ferdinand Dierkens, who was commissioned by Vooruit cooperative. The cooperative defended and united the interests of workers. The cafe-restaurant and the movie theatre provided recreation to the workers and the library was there to develop their intellectual capacities. The aim was to provide luxurious space to the workers and to treat them to the activities that would otherwise be unreachable for them.

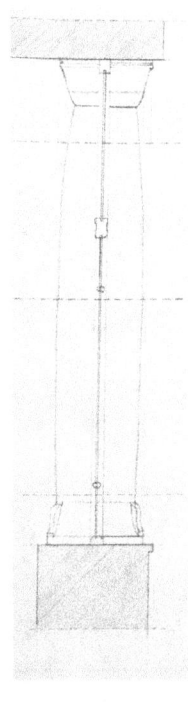

Group 02

Zebrastraat
Private Building
Corner House G-S
Design museum Gent
Studenthome
Private House
Florist's
Twiggy store
Shelter 7
Xi / Café Costume
Stam

BUILDING: Zebrastraat
ARCHITECT: Charles van Rysselberghe
REALISATION: 2010
LOCATION: Zebrastraat 32 Gent
STUDENT: Ban Ngoc Tran

Zebrastraat is an social apartment building, built at the beginning of 20th century and designed by Charles van Rysselberghe. In an oval shape, this building surrounds a courtyard to create a cozy space. From the courtyard, an impressive visual effect could be experienced by such a long curved facade, covered by early 20th century model windows. In this case, the windows for middle class housing were applied to a social project to show somehow a respect to laboring class at that moment. Particularly, in staircase landing, a typical window combines with the curved spatial shape, corresponding to our movement trajectory inside, to enable a feeling of consecutive views down to the nice oval courtyard.

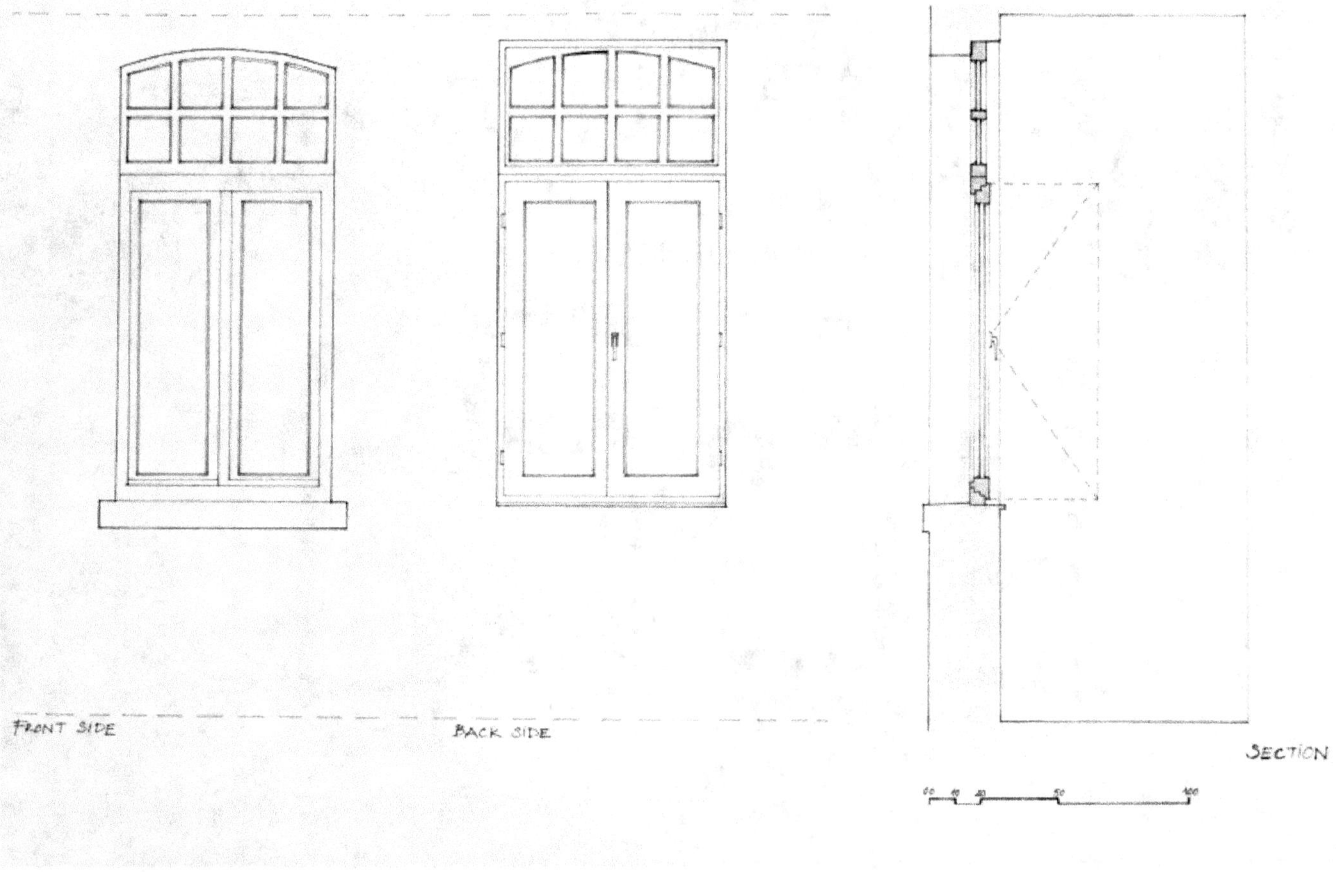

FRONT SIDE   BACK SIDE   SECTION

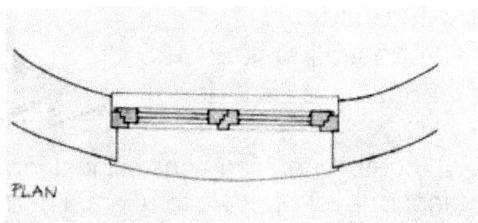

PLAN

23

BUILDING: Private Building
LOCATION: Tentoonstellingslaan 125
STUDENT: Dalia Suzanne Brandone

The building is located in Tentoonstellingslaan n°125, on the second level of it. Victor Compijn, a city engineer, took the initiative for a star-shaped street pattern around a circular plaza. Tentoonstellingslaan is one of the seven streets that arrive to this circular plaza. This avenue has been realized in 1908, on the vacant lands. It was made to smooth the traffic flow in the area, which is the case. Most of the valuable buildings, on Tentoonstellingslaan, came about just before World War I or during interwar period (20's – 30's). These windows give directly on this circular plaza (or roundabout) and what make them particular is that they give a large panorama for the apartment. Indeed, other buildings around this circular plaza don't have this large possible view. Moreover, we can think that living just on a roundabout makes noise, but here it isn't the case, so the aim of the city engineer has succeeded.

ELEVATION scale 1:10

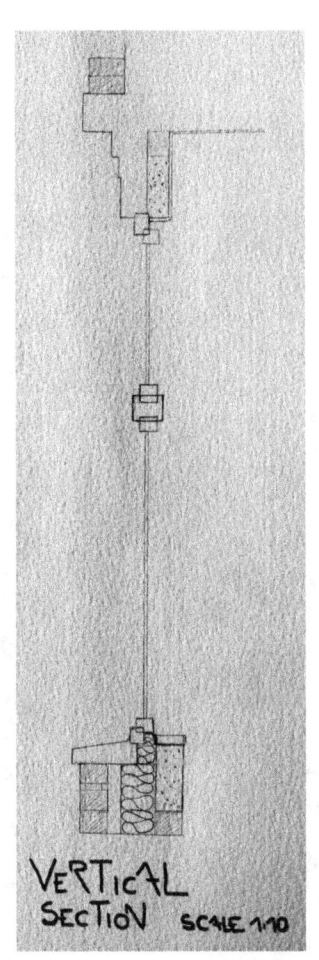
VERTICAL SECTION scale 1:10

HORIZONTAL SECTION scale 1:10

BUILDING: Corner House G-S
ARCHITECT: Graux & Baeyens
REALISATION: 2008-2011
LOCATION: Sassekaai Gent
STUDENT: Jakub Senkowski

This window is located in House G-S in Gent, Belgium. It appears to be an entrance door to 19th century corner house, located at the Muide waterfront area. Great sequence of views from house on the old city harbor docks of Gent, makes this design unique. The idea was conserved the façade, creating a coat to envelop the new spaces. Very large glazing, comparing to the volume of the building provides great connection between inside and outside. Minimalistic black frame and white interior with non-parallel walls are creating very interesting but also intriguing view.

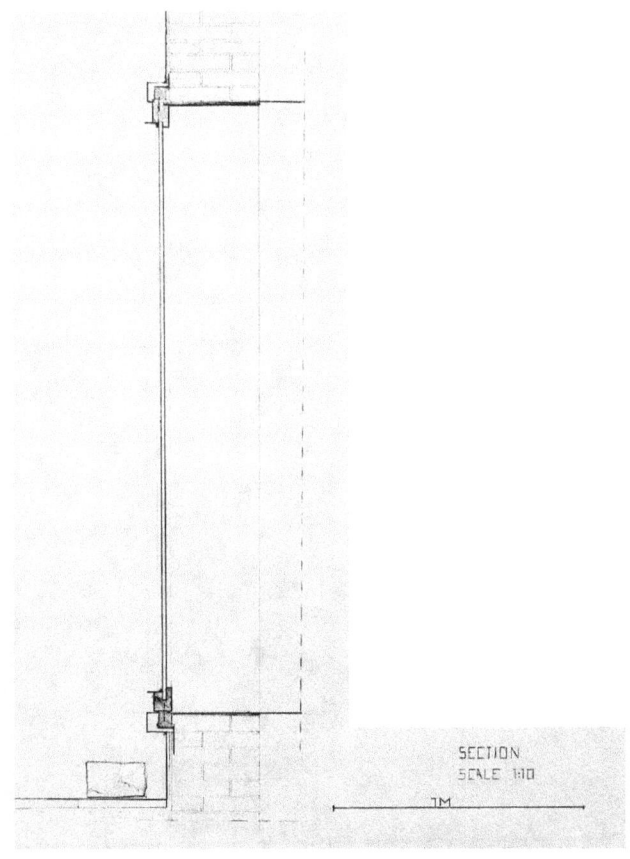

SECTION
SCALE 1:10

1M

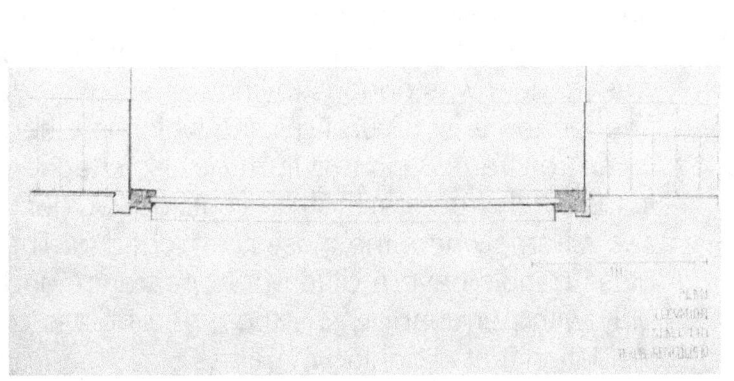

BUILDING: Design Museum Gent
ARCHITECT:
REALISATION: 1675
LOCATION: Jan Breydelstraat 5
STUDENT: Jolien Vandereecken

I chose this project because, not for it's appearing but I walked by a few times and asked to myself how the building and the glass façade was made. I was curious to know.
The facade is structural glazing which is glued on an iron frame, and that frame is also attached to the steel structure. From the outside you don't see a frame where the glass is attached to. The glass panels have a different appearance and play with permeability, by the type of glass or insulation that is behind the glass.

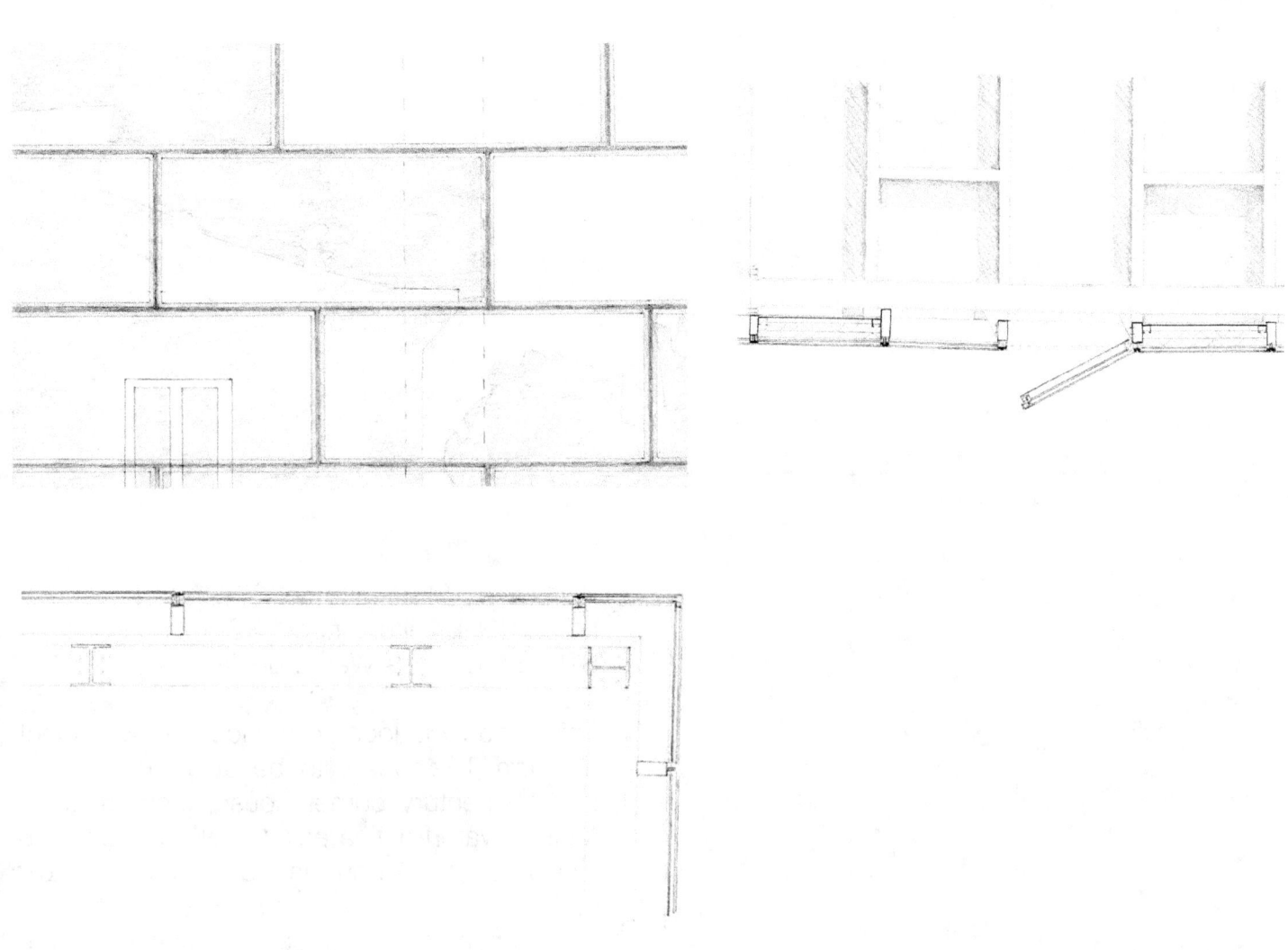

BUILDING: Studenthome
LOCATION: Pekelharing 14 Gent
STUDENT: Jonas Werkbrouck

This window is located in House G-S in Gent, Belgium. It appears to be an entrance door to 19th century corner house, located at the Muide waterfront area. Great sequence of views from house on the old city harbor docks of Gent, makes this design unique. The idea was conserved the façade, creating coat to envelop the new spaces. Very large glazing, comparing to the volume of the building provides great connection between inside and outside. Minimalistic black frame and white interior with non-parallel walls are creating very interesting but also intriguing view.

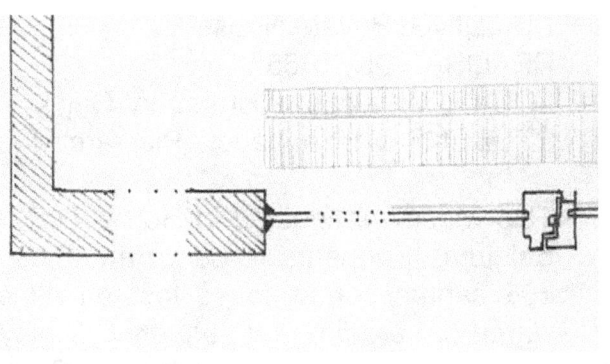

BUILDING: Private House
REALISATION: 1655
LOCATION: Zwartezusterstraat Gent
STUDENT: Marika Izabela Pierkarczyk

This window is located in House G-S in Gent, Belgium. It appears to be an entrance door to 19th century corner house, located at the Muide waterfront area. Great sequence of views from house on the old city harbor docks of Gent, makes this design unique. The idea was conserved the façade, creating coat to envelop the new spaces. Very large glazing, comparing to the volume of the building provides great connection between inside and outside. Minimalistic black frame and white interior with non-parallel walls are creating very interesting but also intriguing view.

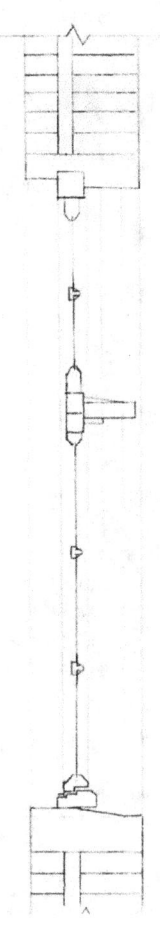
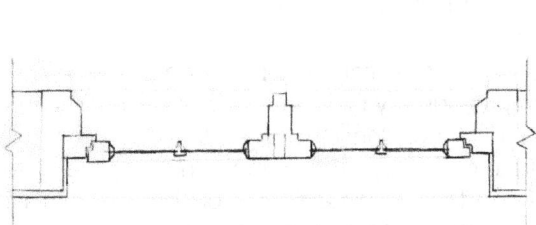

BUILDING: Florist's
ARCHITECT: Van der Weken
REALISATION: 2013
LOCATION: Citadellaan 4 Gent
STUDENT: Martin Mikovcák

I stopped, my attention was focus on that window, in first view simply, primary but something is not correct. Where is that window ? Suddenly i realised that i can see just frame, behind them is green hole into the house , there is no garden there is florist ! Frame is there but window is missing.The frame includes front massive gate, which show intractability, but they can be open and hide behind the wall and we can see main entrance to the florist which is inviting us inside. But there are one more doors hidden into the wall, private doors which looks untouchable. When we will enter to the florist we can see the missing window hidden behind the edge. Openness give us freedom we have feeling of nature but also we can feel context of the street, which is touchable near and without any bariers. But still we can feel save, because not many attention is focused on us, the frame is easy to miss in the surrouding. And now i know what foccused my attention, relationship to the surrounding, new frame in old building plugged in with patience to consciousness of sensityvity.

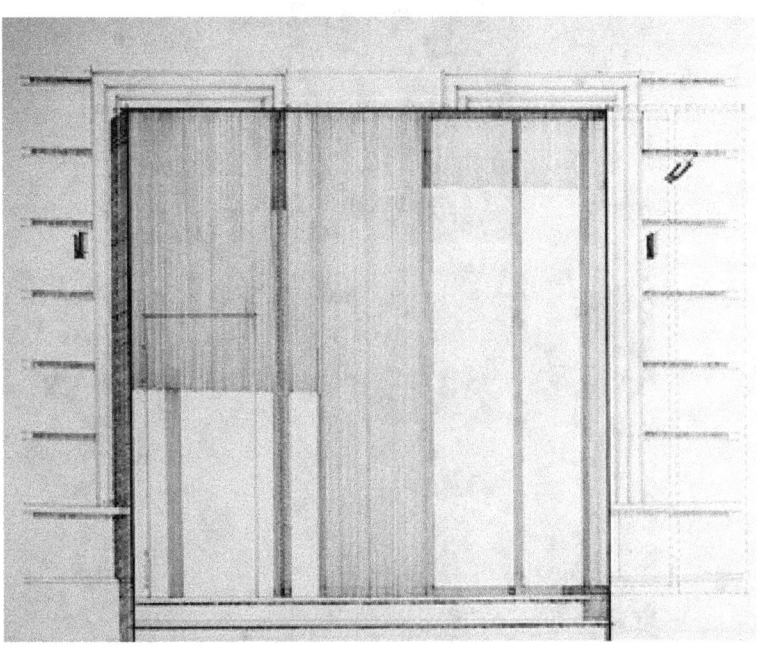
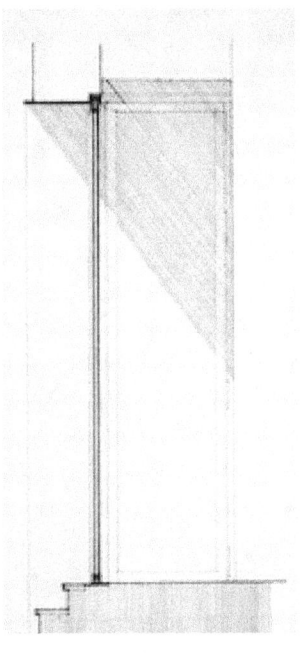

BUILDING: Twiggy Store
ARCHITECT: De Vylder Vinck Taillieu
REALISATION: 2011
LOCATION: Notarisstraat 3 Gent
STUDENT: Petra Ross

Is it necessary for historic windows to stay always symmetric? No, it doesn't ...
The architects kept the old house as a monument and added little abstract touches seen only on the second view. Playing with the stairs, the windows, mirroring them and using the old damaged walls is part of the game. Contrast of old and new creates the final timeless look perfect for a clothing shop. I like timeless stuff and spaces which can surprise you. I consider this building as a nice sensitive example of contemporary reconstruction of the 19th century townhouse.

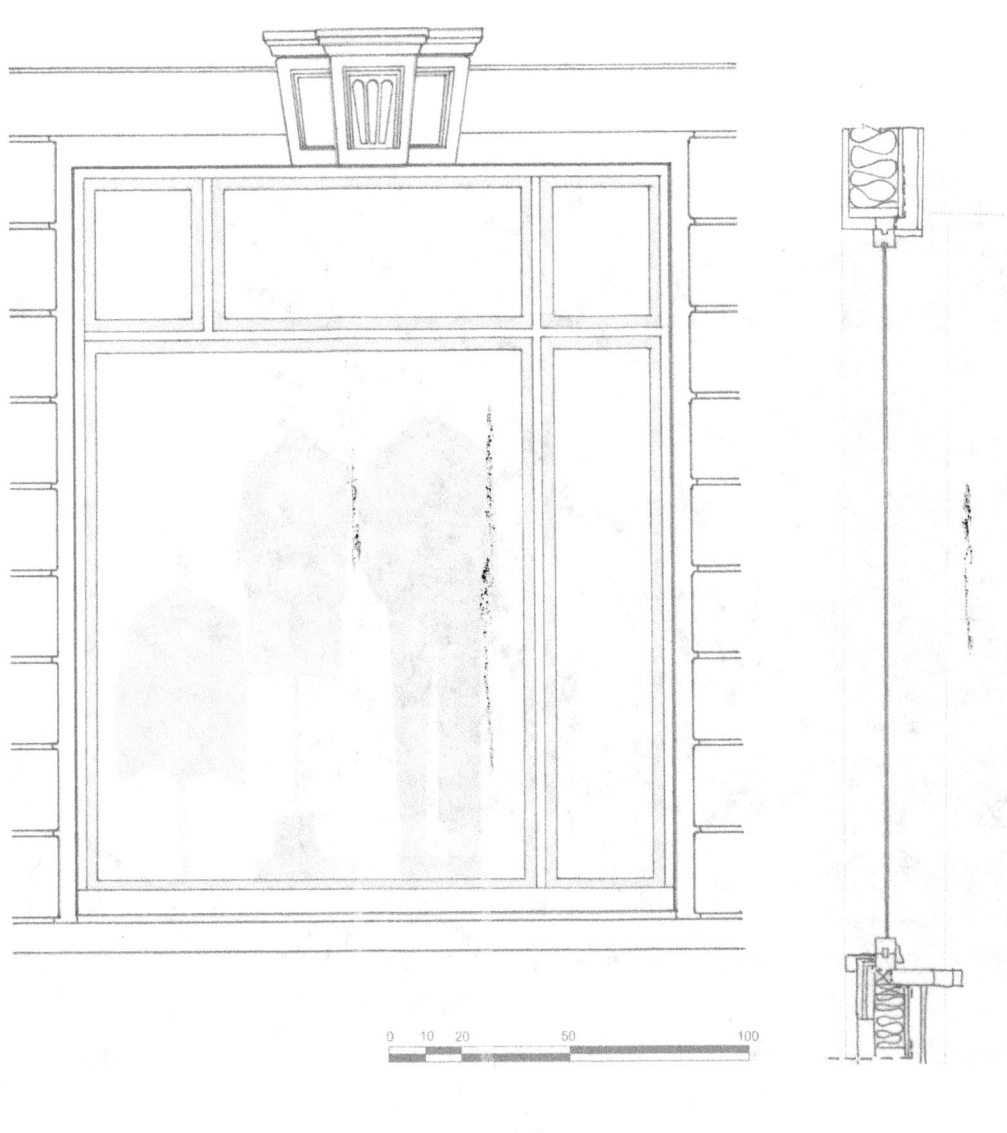

BUILDING: Shelter 7
LOCATION: Bennesteeg 7
STUDENT: Radka Vilimkova

This window is hidden treasure, it's behind thick wall and you can see just a little part of it. When you'll go further inside you'll find it. Nearly 9 meters high glass wall made from three glass pieces one above each other. There is no visible frame so it looks really pure.
Inside there's bike shop and Shelter 7, which is design hotel, an inspiring space serving as a gallery, shoot location, brainstorm room, meeting point or even pop-up store.

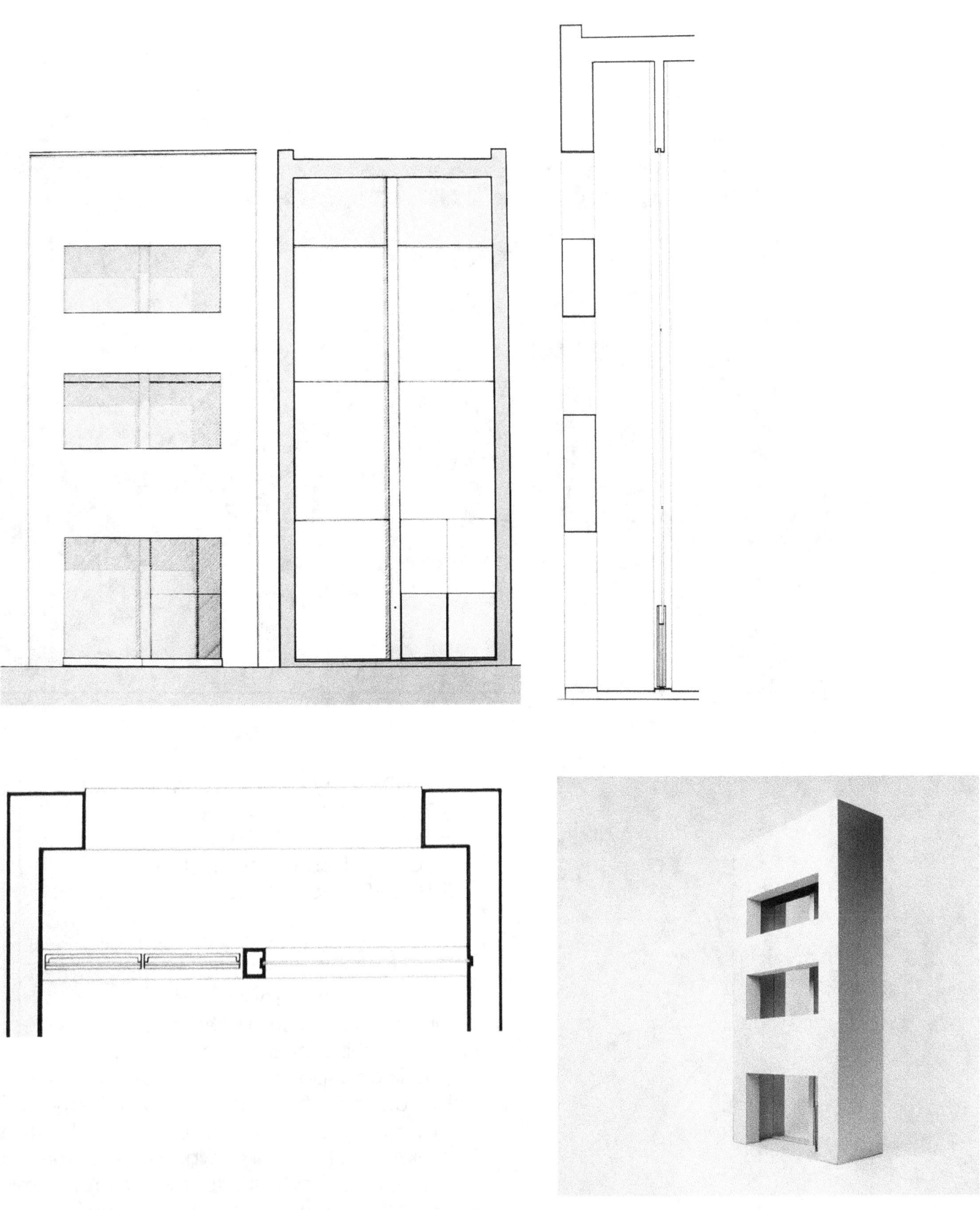

BUILDING: Xi / Café Costume
ARCHITECT: Louis Cloquet
REALISATION: 19th Century
LOCATION: Brabantdam 135 Gent
STUDENT: Stepan Vasut

During a walk in Gent we found a Red Light District, which attract us to go in. After we had went throw windows with girls behind we found small church-square, where exactly this window was. I really liked it.
I get in and spoke to a owner of a tailor shop. He told me, that this window is pretty rare because it appeared in a movie called The broken circle breakdown, which earned a place at Oscar's. It's funny, that randomly picked window can be that famous.

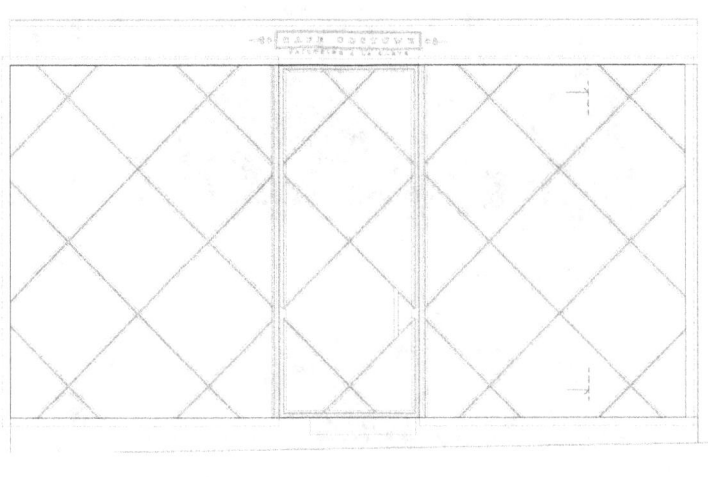

BUILDING: STAM
LOCATION: Godshuizenlaan 2 Gent
STUDENT: Tomas Matata

Choosing the window, I wanted relation of a historic and modern architecture. Beside the robustness of the Stadsmuseum's entrance hall, there's a small building with function of a sheltered picnic place. It's a historic building adjusted to its new purpose. The front facade is entirely made by glass as a modern addition. Doors are installed a bit before the facade, covering the brickwork with glass. Through the window it's possible to see the wooden structure of the roof. This intervention is not trying to get attention and overshadow the historic part, on the contrary, it's supporting and promoting it.

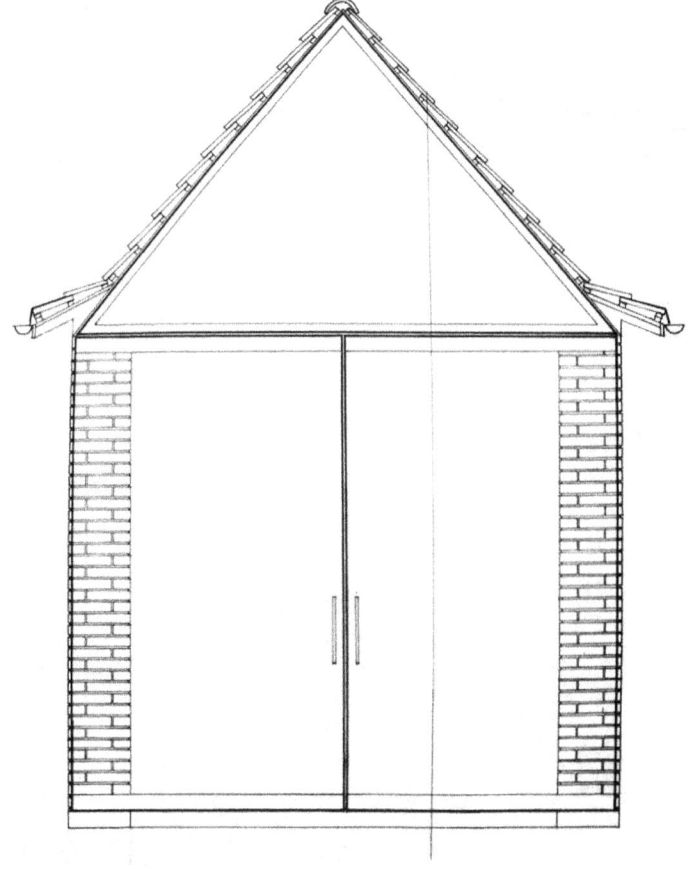
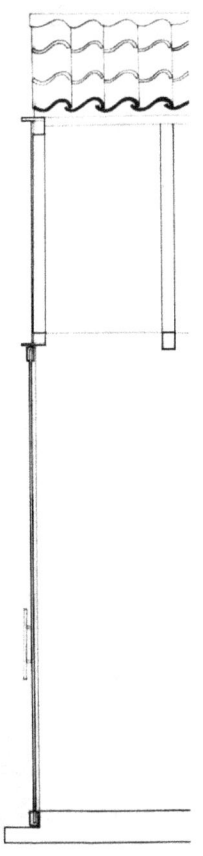
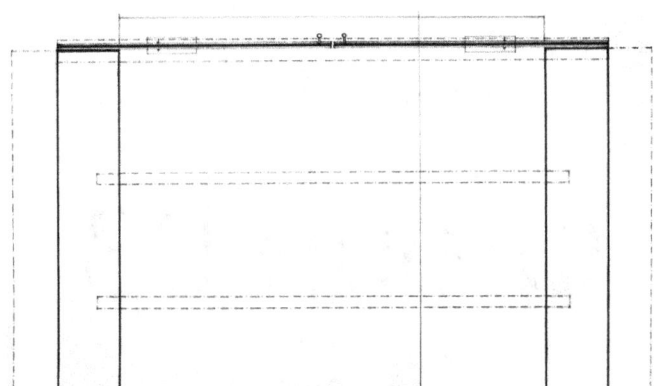

Group 03

Ramen

Private House

Bakerij Himschoot

HoGent-Bijloke Campus

Masons' Guild Hall

Gravensteen Castle

Pakhuis restaurant

The House of Alijn

Campus KTA Casinoplein

Les Ballets C deB and LOD

BUILDING: Ramen
ARCHITECT : Els CLaessens and Tania Vandenbussche
REALISATION: 2007
LOCATION: Ramen, Gent
STUDENT: Anouk Desmedt

The project is called Ramen, it's a residential building located in the center of Gent. The masterplan is made by Hilde Huyghe and Tomas Nollet and the different housing units are done by different architects. In workshops the different architects came toghether to make a coherision between the diffrents facades. The architects of the facade this window is comming from are Els Claessens and Tania Vandenbussche. The window has this kind of duality between the window and the ballustrade wich makes it interesting. The windows are really high and you can open them completely this gives you a good relation with the garden and the interior of the room.

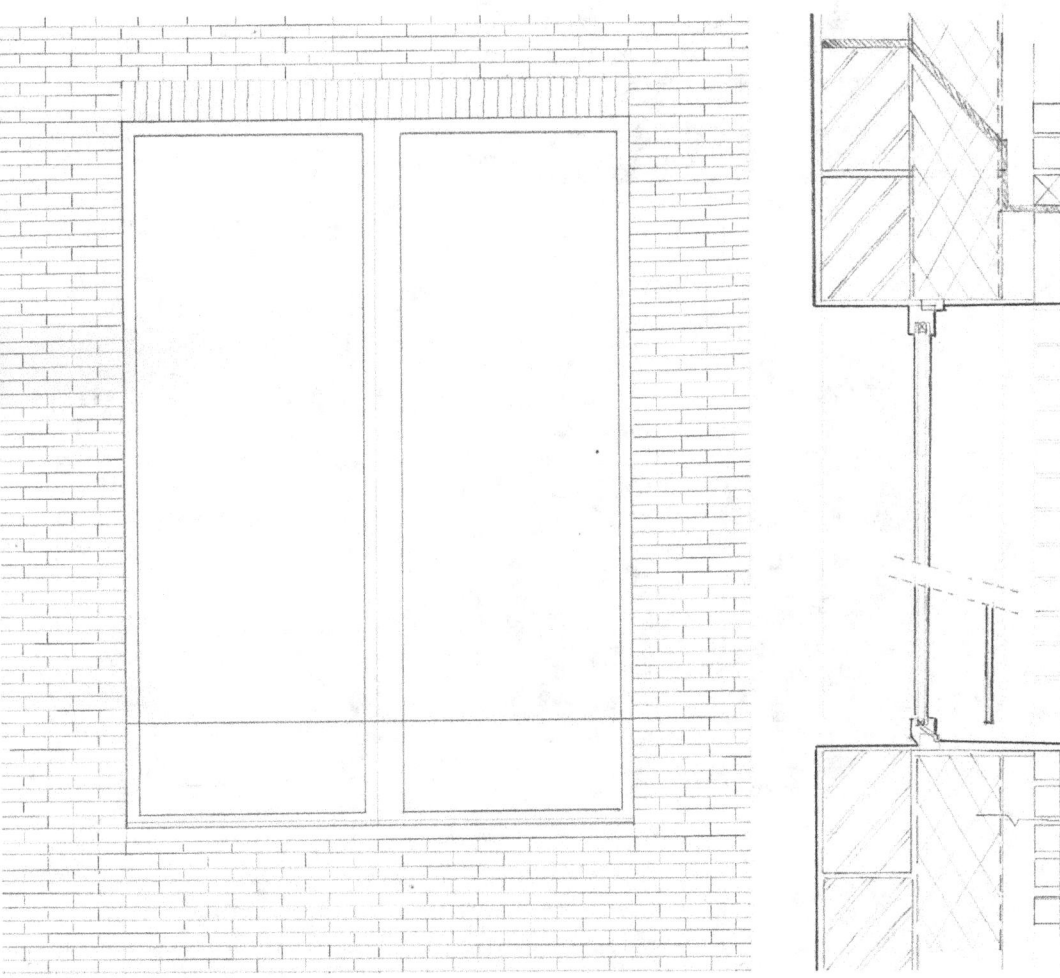
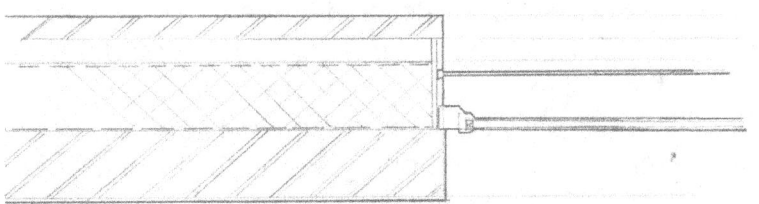

BUILDING: Private house
REALISATION: 14th Century
LOCATION: Nodenaysteeg 4, Gent
STUDENT: Dang Phuong Dan Tran

The chosen building is a private house located on Nodenaysteeg Street. This street has a long time history, it belonged to the Nodenay family in 1357. The house was also constructed at that time, after hundreds year and suffering from war, it still remains until now. The house was made from stone and the structure is maintained as its original.

This house is located in a small alley within the center of Gent. Although the vibrant and dynamic activities surrounding, the alley and the house bring the feeling of the past which is peaceful and idyllic. The two layers window which are made from wood and glass have a strong identity of Gent in the past.

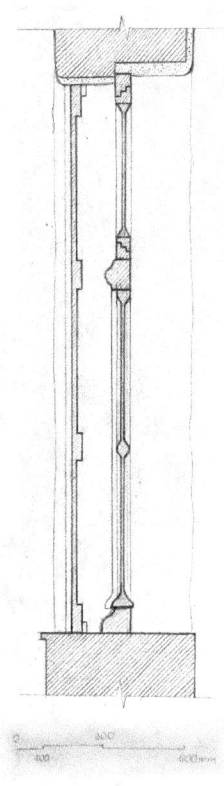

Preville house
vaanaysteeg 4, Gent

BUILDING: Bakerij Himschoot
REALISATION: 17th Century
LOCATION: Groentenmarkt 1 Gent
STUDENT: Erika Vandekerckhove

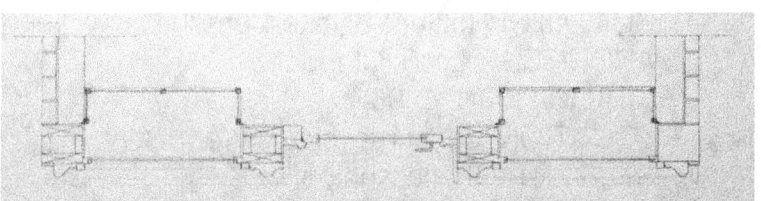

BUILDING: HoGent-Bijloke Campus
ARCHITECT: Louis Cloquet
REALISATION: 1896
LOCATION: Jozef Kleyskensstraat Gent
STUDENT: Ghaith Khalouf

Choosing Bijloke Campus window was due its nobility and its historical and cultural value. The building built in 1896 and it is still in good condition. The windows are all stony decorated with wooden frame. The building is currently used as a school campus which reflects the cultural Impact it has. The large windows allow sunlight to presence in the building offering optimum natural lightening inside.

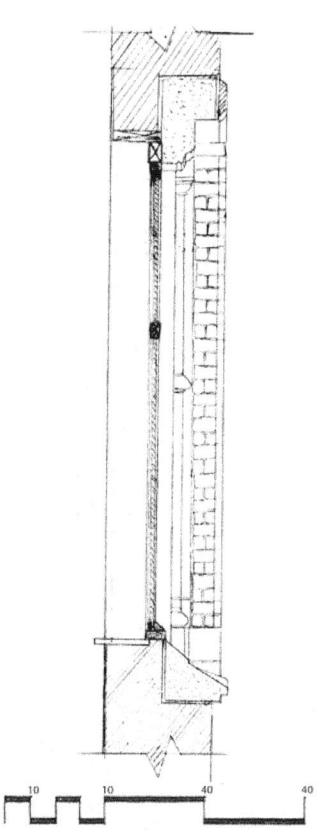

BUILDING: Masons' Guild Hall
REALISATION: 16th Century
LOCATION: Sint-Niklaasstraat 2
STUDENT: Kaitaro Hirai

Opposite St Nicholas' Church, on the other side of the street, is the original 16th-century Masons' Guild Hall. In this architecture, I think the frame of window is sharp and thin than that of typical old building in Gent. The frame has the exact vertical and horizontal line. Especially the character of this window is the layer of it. There are two layers of the frame and glass. One the side is covered by glass and the other one is not. Behind the front window, there is a window used as it usual.

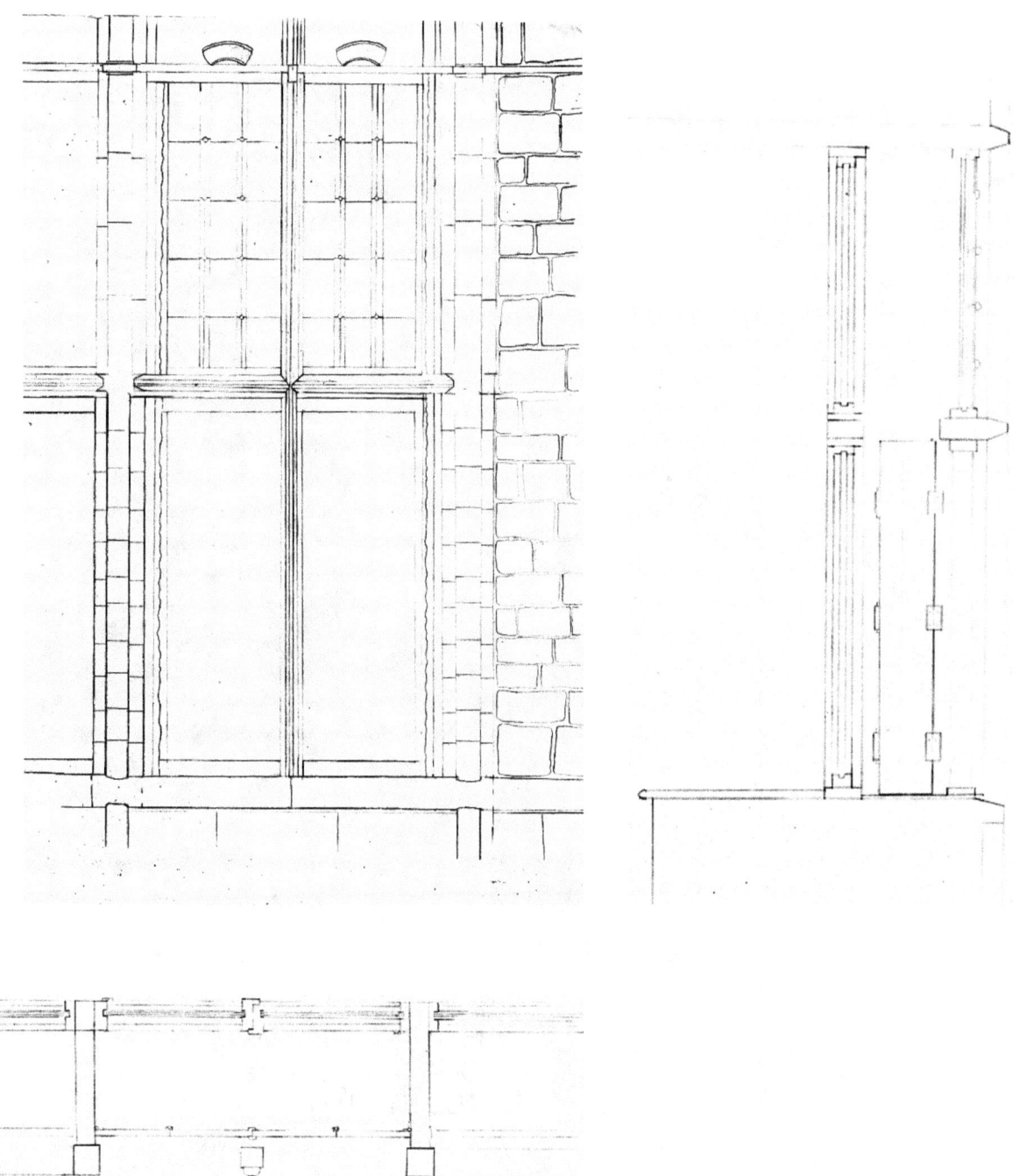

BUILDING: Gravensteen Castle
REALISATION: 1600
LOCATION: Sint-Veerleplein 11 Gent
STUDENT: Thi Khanh An Phan

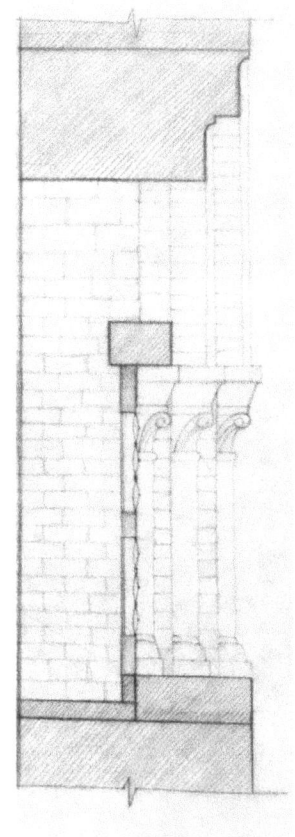

BUILDING: Pakhuis Restaurant
ARCHITECT: Antoine Pinto
REALISATION: 1989
LOCATION: Schuurkenstraat 4 Gent
STUDENT: Minh Ha Nguyen

Pakhuis is a restaurant located within the historical district. Being constructed in 1989 and designed by Antoine Pinto, the architect try to bring the image of industrialize area with metal structure into the center of Gent where most of the construction are in stone and brick. I chose the window due to its unique and contrary within the building's context. Lying on a brick arc in a small alley, the window show its contrary with a different look with glass and metal. More than that, it also reflects and reminds visitors a sense of a passed industrial area.

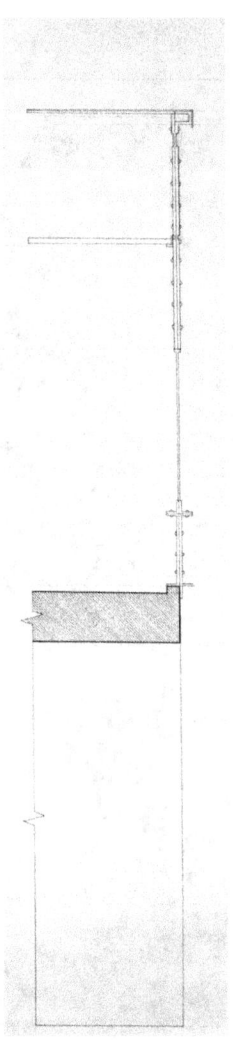
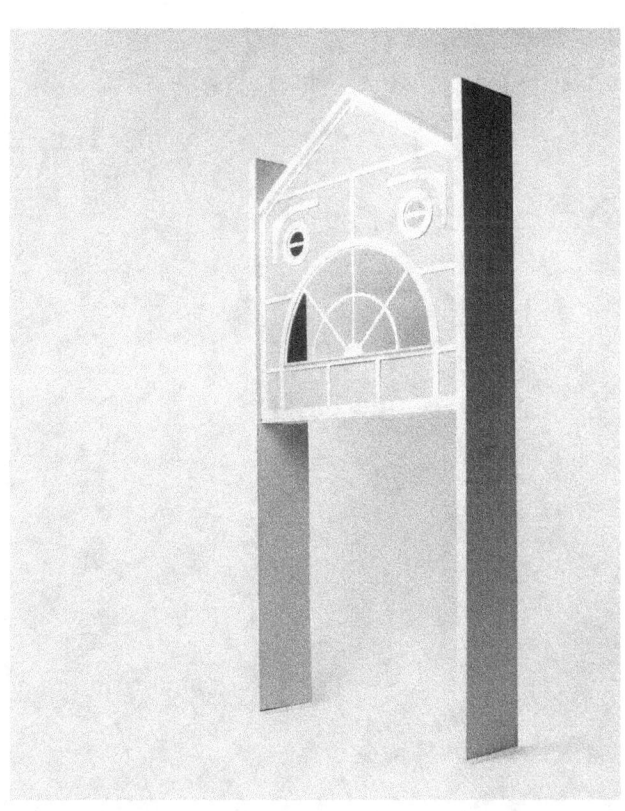

BUILDING: The House of Alijn
REALISATION: 1363-1546
LOCATION: Kraanlei 65, Gent
STUDENT: Oanh Nguyen Thi Kieu

The story of "The house of Alijn's window" is told by century to century, go through variant kinds of architectural functions, from the almshouse, pediatric hospital to the workshop, and ended-up as the museum. Affected by the low technology in construction, also the low charity budget for building this complex, the window has a humble vision and condensed pattern. This visualization brings the natural light to the interior room in a charming way that can create the cozy atmosphere inside.

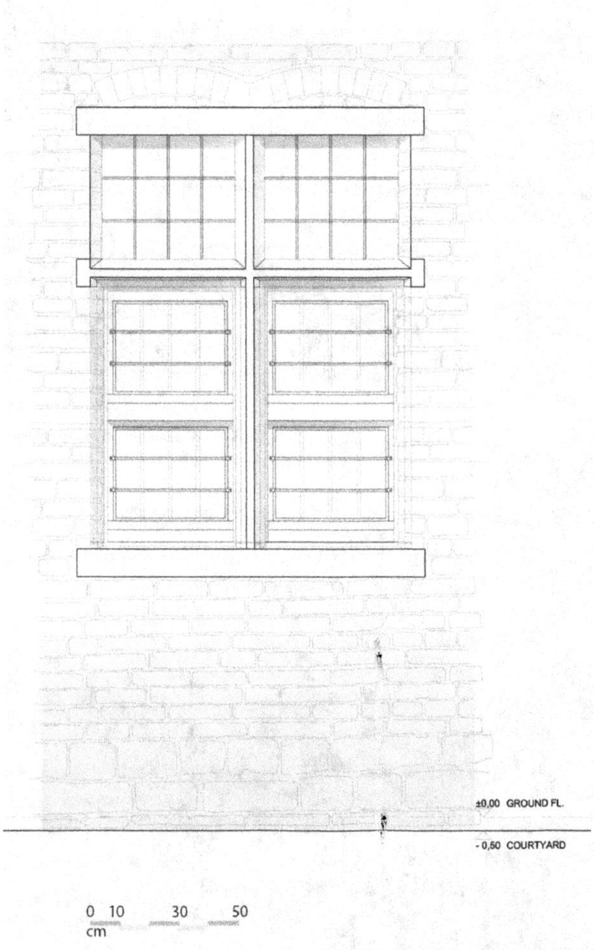

±0,00 GROUND FL.
- 0,50 COURTYARD
0 10 30 50
cm

CH +2,57
±0,00 GROUND FL.
0 10 30 50
cm

CH +2,5
±0,00

0 10 30 50
cm

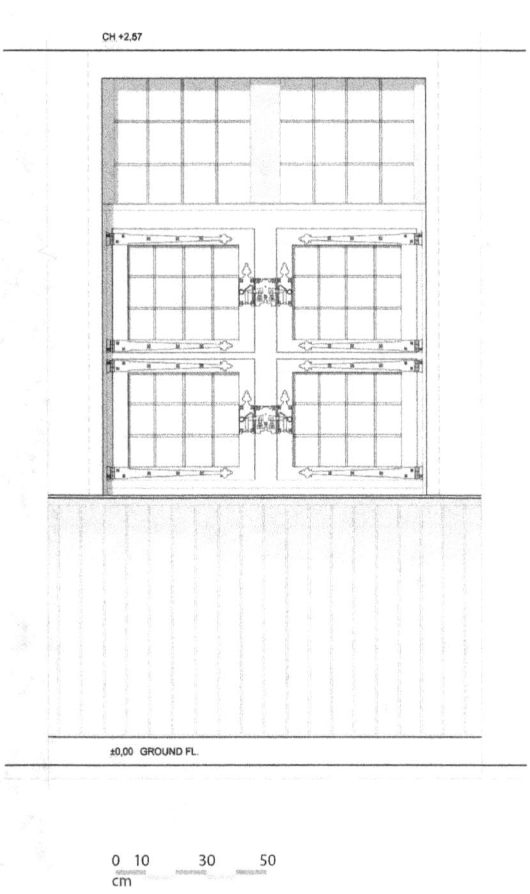

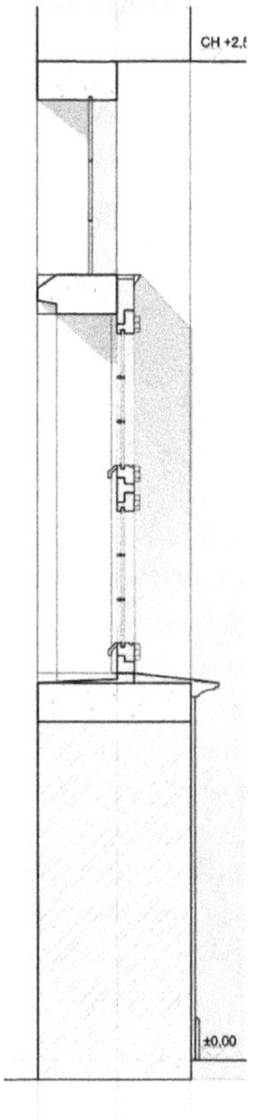

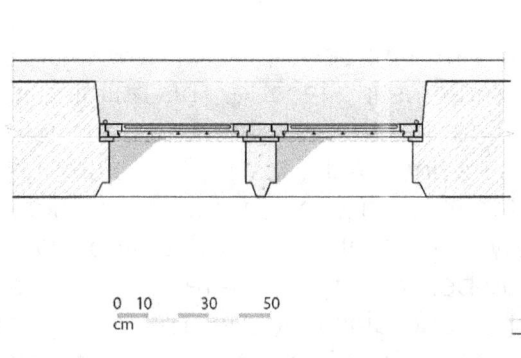

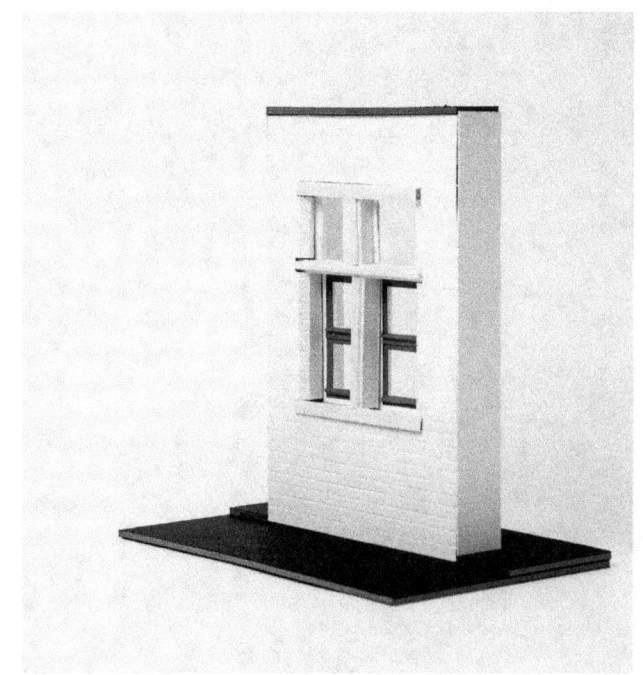

BUILDING: Campus KTA Casinoplein
ARCHITECT: Abscis Architecten
REALISATION: 2012
LOCATION: Coupure Rechts 312 Gent
STUDENT: Takeshi Uchida

The reason why I choose this window is this building has interesting façade. This building is located beside the channel, and there are lots of buildings having brick façade. Therefore the choice of façade, brick or different new façade, is really important. Abscis Architect who won the competition chose interesting façade between brick and glass façade. That is 'Glass Brick Façade'. In addition, this building can keep the harmony by the glass brick which looks like brick and clear reflection which come from smooth glass line.

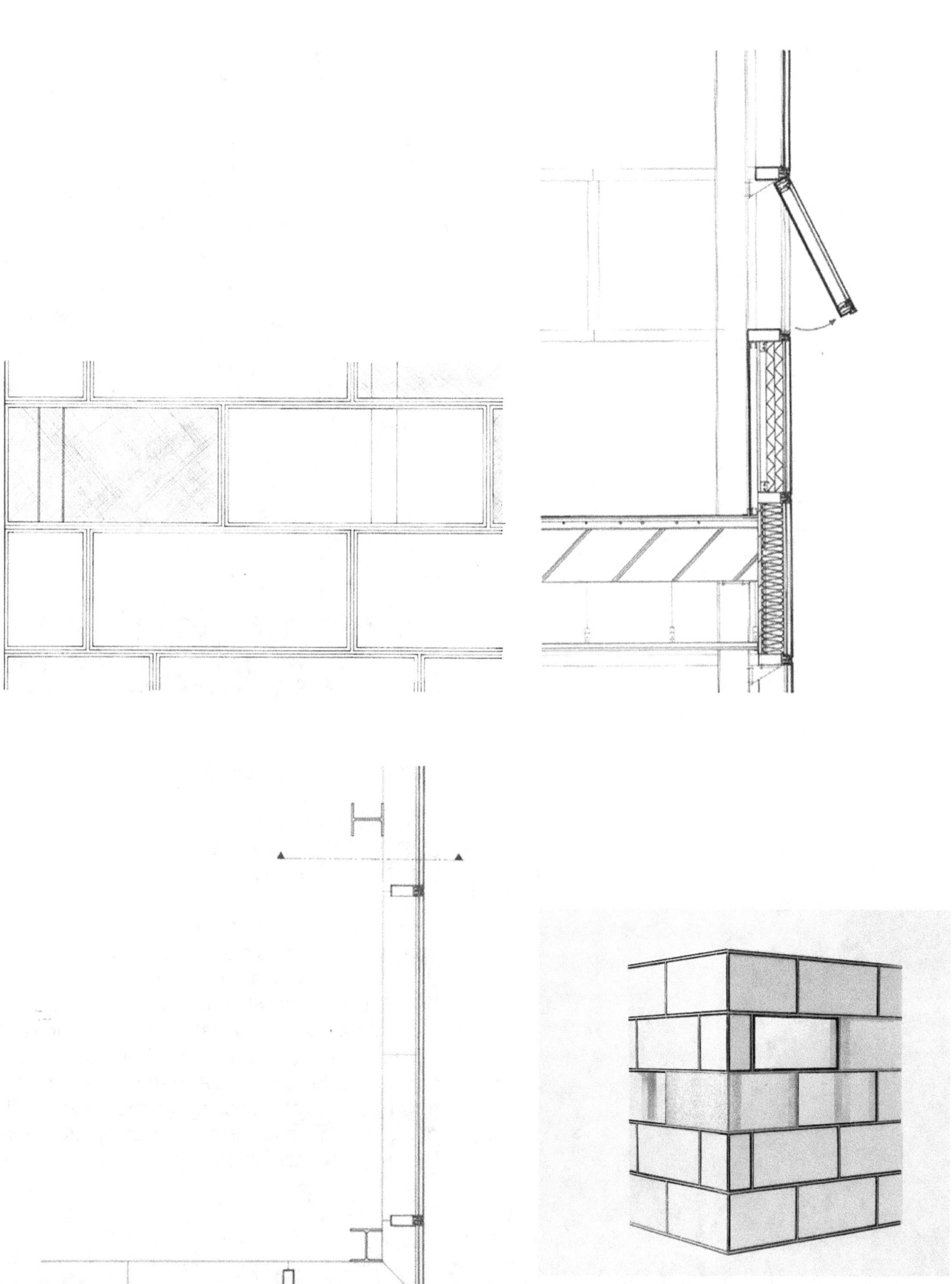

63

BUILDING: Les Ballets C deB and LOD
ARCHITECT: Jan De Vylder
REALISATION: 2008
LOCATION: bijlokekaai Gent
STUDENT: Yuki Fujita

There are two identic buildings on this site. The open facade shows the construction of this building. We can look at the detail of columns and beams or materials.In spite of such as this curtain wall, some windows are put again. They are for not only showing the detail of building but also refreshing the airs. And they look like mirrors because of reflecting neighborhood buildings.Outside and inside sceneries are overlapping with each other.

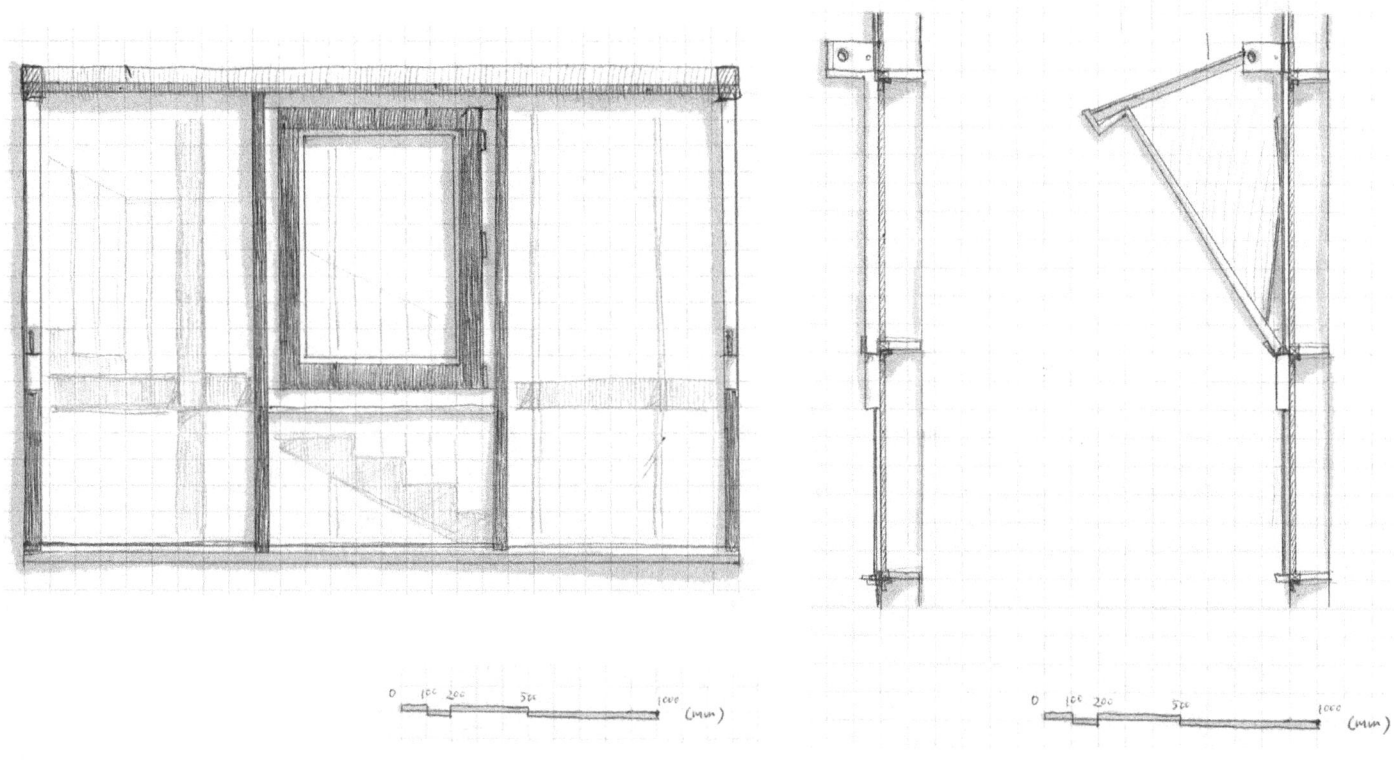
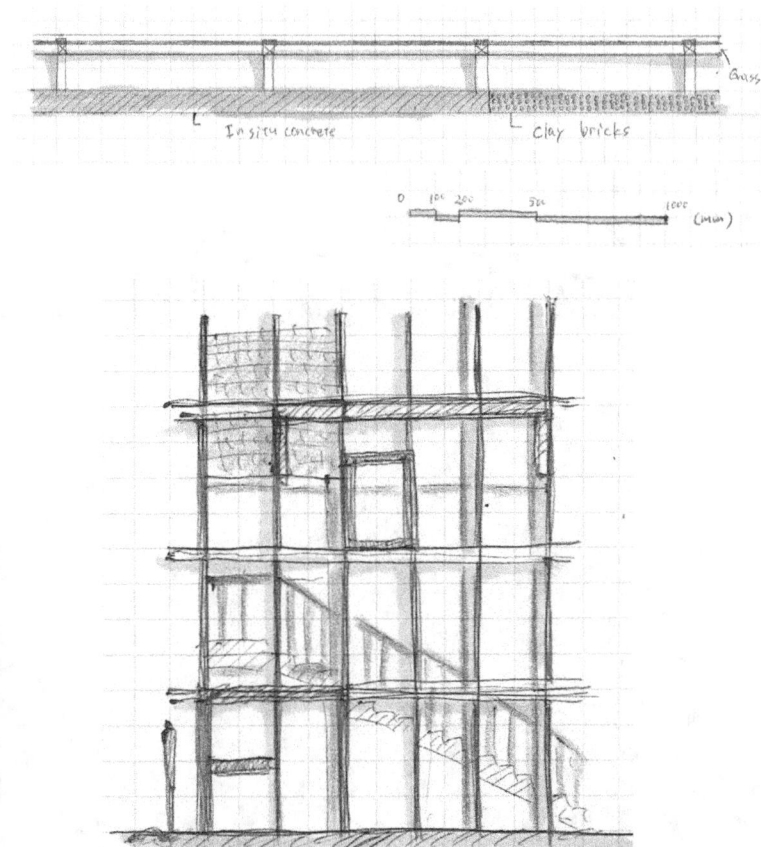

Glass
In situ concrete   Clay bricks

Group 04

Vlaamse Opera
Artevelde Hogeschool Kantienberg
Hotel Falligan / Literary circle
De Oudburger
Stadhuis
Private House

BUILDING: Vlaamse opera
ARCHITECT: Bernard de Wilde
REALISATION: 1737
LOCATION: Schouwburgstraat Gent
STUDENT: Alessandro Martinelli

The theater of the archers' guild of Saint Sebastian in the Coulter was build in Gent in 1698 and made it suitable for opera performances. The building was inaugurated with the staging of the opera ThÈsÈe by the French composer Jean-Baptiste Lully by the group of Giovanni Paolo Bombarda and Pietro Antonio Fiocco, who founded La Monnaie in Brussels. This building burned down in 1715. The building has been renovated between 1837-1840 in neo-classical style, by the then town of Gentís architect Louis Roelandt on the site of the demolished St. Sebastian Theatre.

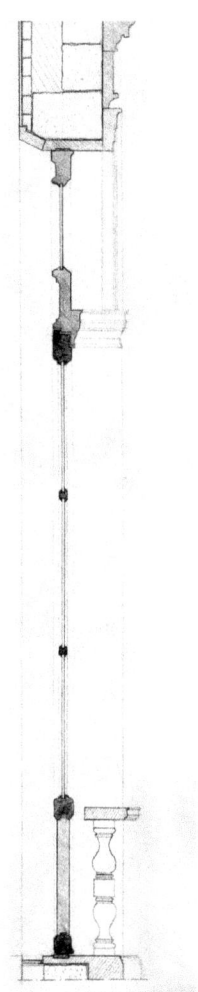
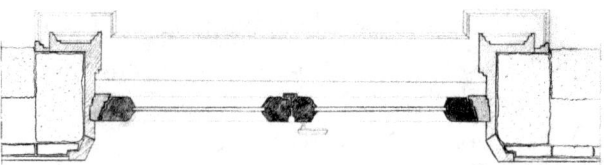

BUILDING: Artevelde Hogeschool Kantienberg
ARCHITECT: Crepain Binst Architecture nv
REALISATION: 2000
LOCATION: Site Overpoort Gent
STUDENT: Ilona Kerstin Pietrus

WHY?
I've chos en this building, because it was the first building, which appearance impressed me at my first day in Gent.It combines current architecture with local identity.
The chosen Panorama window gives the imagination of beeing part of the outside, so it acts like a bridge from inside to outside and vice versa.
WHAT?
The designers of this building focussed on the experience of LIGHT , AIR , SIMPLICITY and BEAUTY, which you also can see in the construction of window.
WISE...
The school was built in order to collect the students of four catholic schools with in total 6700 students. Nowadays with 11000 students its the biggest catholic school Flanderns!

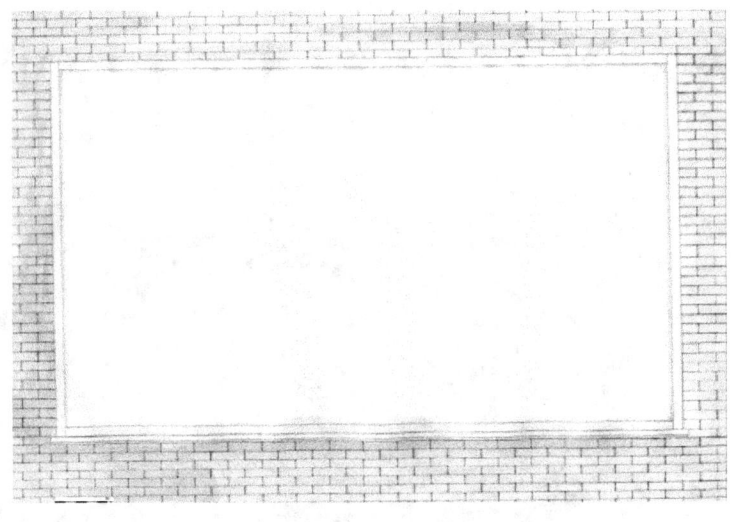
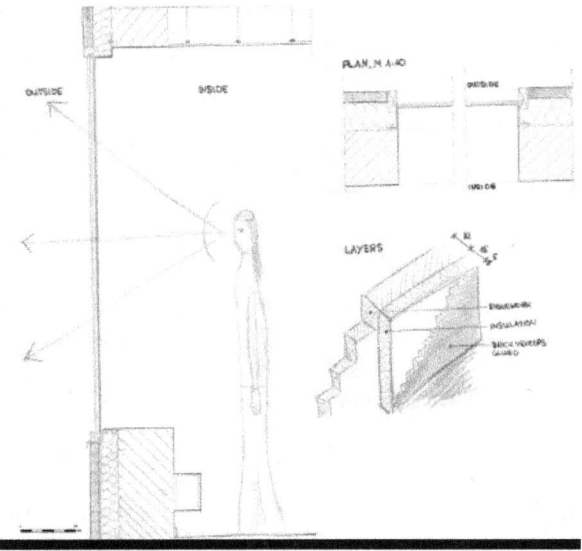

BUILDING: hotel Falligan / Literary circle
REALISATION: 1755
LOCATION: Kouter 172 Gent
STUDENT: Kalin Hristov

This magnificent building may rightly be called the jewel of the Kouter square. The building was erected in 1755 by order of knight Hector Falligan.
The imposing facade of nine bays has a raised middle part topped by a triangular pediment. Freestanding columns
with Corinthian capital emphasize the central section. The two images above the pillars set for Apollo and Diana.
The delicate arched windows on the façade give the building added value. Their framework is made by timber and has a
lot of artistic curves. At first the place was meant to be a hotel, but in the course of the 19th century literary society "Club des Nobles" settled in the building. Even today it meets that association under the name of "literary circle". In 2001 the façade took its original color back.

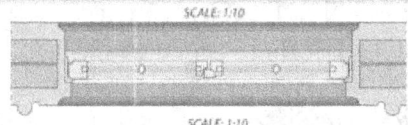

SCALE: 1:10

SCALE: 1:10

BUILDING: De Oudburger
ARCHITECT: Lieve Van De Walle
REALISATION: 1669
LOCATION: Kraanlei/Rodekoning
STUDENT: Kateryna Taratynska

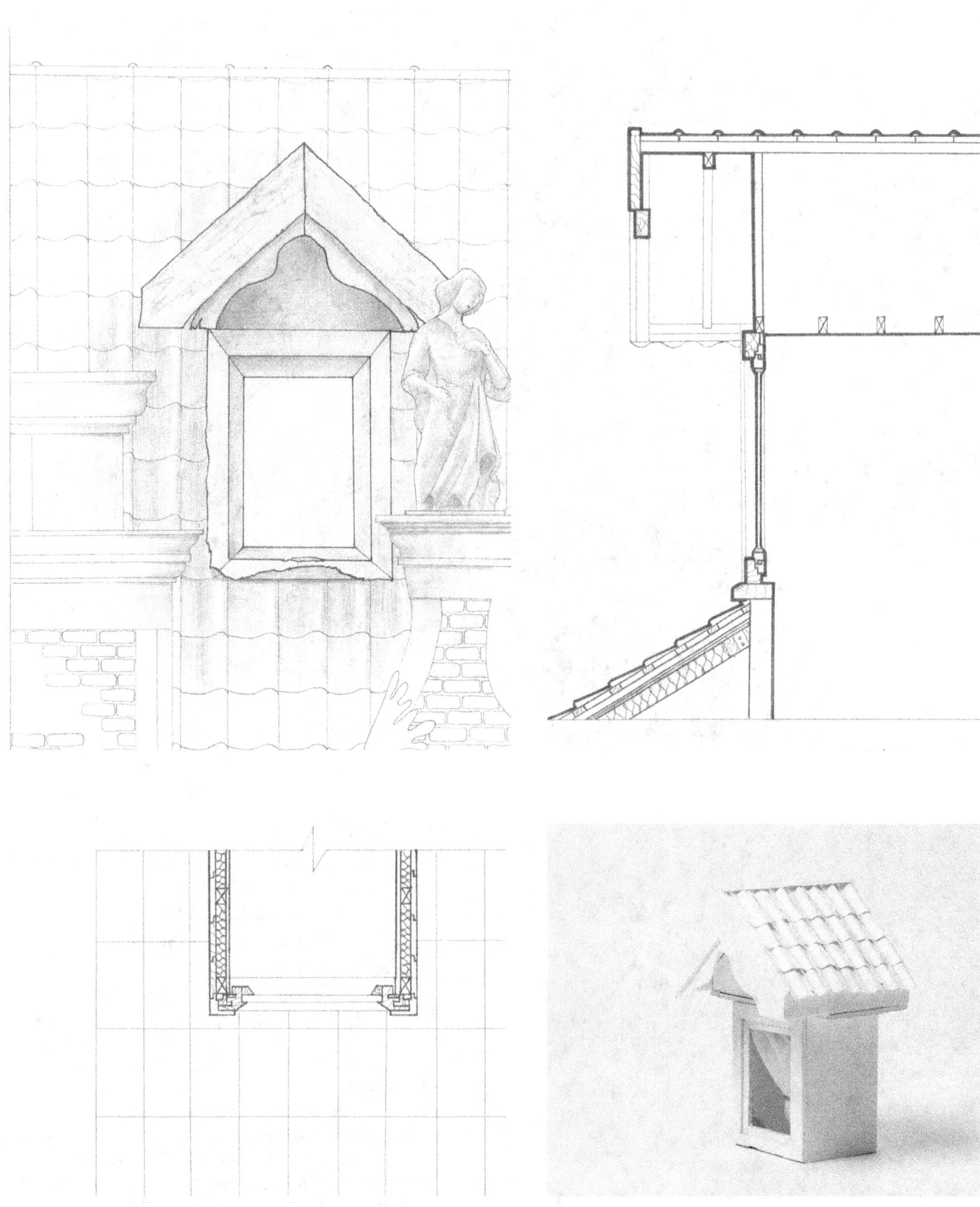

BUILDING: Stadhuis
ARCHITECT: Rombout II Keldermand and Dominicus de Waeghemaekere
REALISATION: Renassance
LOCATION: Botermarkt 1 Gent
STUDENT: Mandana Fouladi

Stadhuis is an early example of what raising taxes can do to a city. In 1516 Antwerp's Domien de Waghemakere and Mechelen's Rombout Keldermans, two prominent architects, were called in to build a town hall that would put all others to shame. However, before the building could be completed, Emperor Charles V imposed new taxes that drained the city's resources. The architecture thus reflects the changing fortunes of Gent: the side built in 1595-6018 and facing Hoogpoort is in flamboyant Gothic style; when work resumed in 1580, during the short-lived Protestant Republic, the Botermarkt side was completed in a stricter and more economical Renaissance style; and later additions include Baroque and rococo features. The tower on the corner of Hoogpoort and Botermarkt has a balcony specifically built for making announcements and proclamations; lacelike tracery embellishes the exterior.

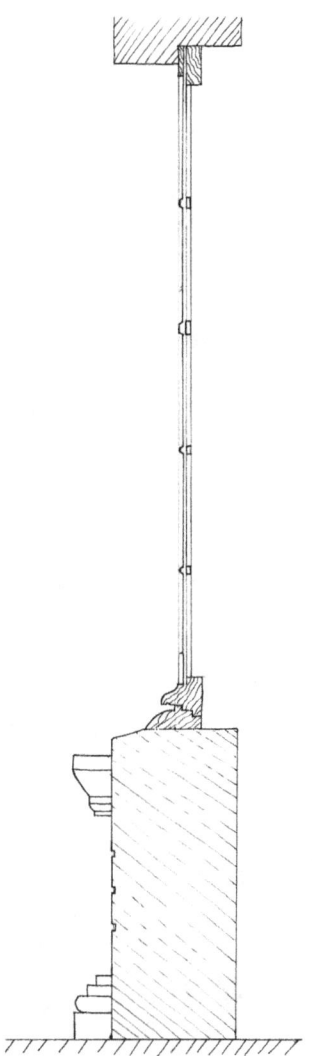
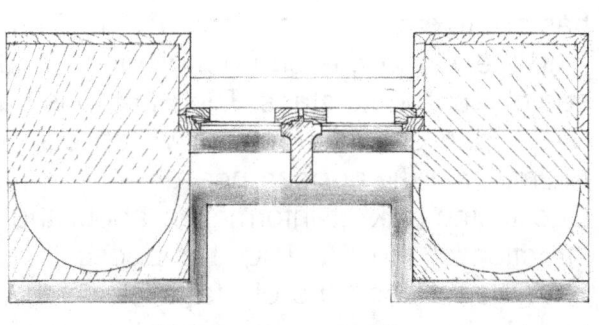

BUILDING: Private House
ARCHITECT: Karel Willems
LOCATION: Wintertuinstraat 29
STUDEN: Zhana Ivanova

This building is a private family house. The architect of the building is a Belgian freelance architect Karel Willems and he lives there with his family.

I chose to represent the big square window. It has approximate dimensions of 2,4 / 2,4m. Behind the window, even from distance, you can see elegant white stairs. The window is framed by welded sheets of metal, which completes the minimalistic appearance.

Due to the ack of information about the construction of the building, I suppose that they have demolished the old façade, which might have looked like the row houses that surround it which are in one and the same style.

ELEVATION

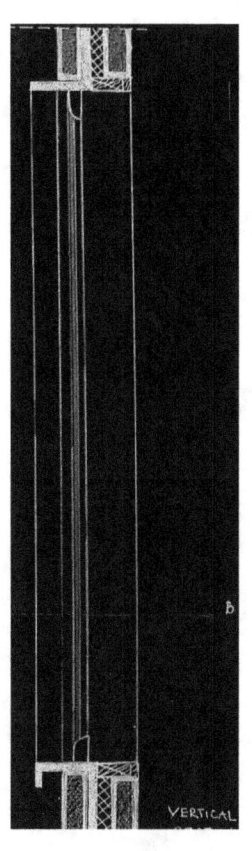
VERTICAL

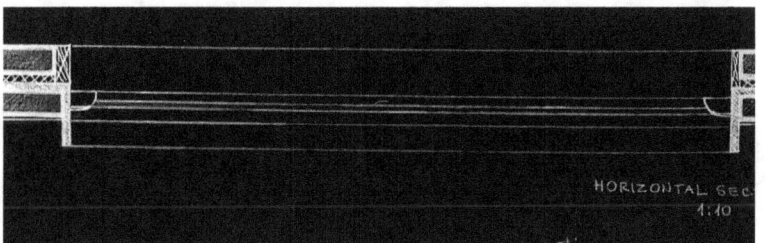
HORIZONTAL SEC.
1:10

Group 05

Booktower
Sint-Barbaracollege
Café het Spijker
Private House
Frank Steyaert's Atelier
Fallen Angles store
Desmet-Guequier cotton mill
Poezie centrum
Nucleo Kunstenaars atelier
Coffee Max
Platte beurs coffee

BUILDING: Book tower
ARCHITECT: Henry Van de Velde
REALISATION: 1942
LOCATION: Rozier Gent
STUDENT: Alejandro Dominguez

In 1933, one of the most well-known Flemish architects, Henry Van de Velde, was commissioned to design a new building for the university library and for the institutes for Art History, Veterinary science and Pharmaceutical science. It was going to be built on one of the highest grounds in the city of Gent. Linking this with the vision which Van de Velde had in mind, storing the books in a tower, the project comes to an icon for the city, a symbol that shows the knowledge, the university. This is why it is really interesting the meaning of every window in the Book Tower, what they look at and what they store.

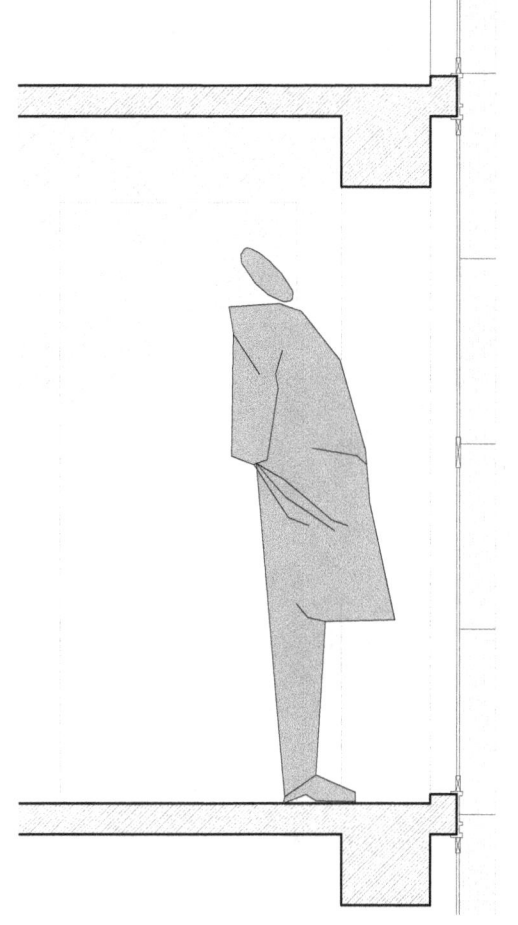
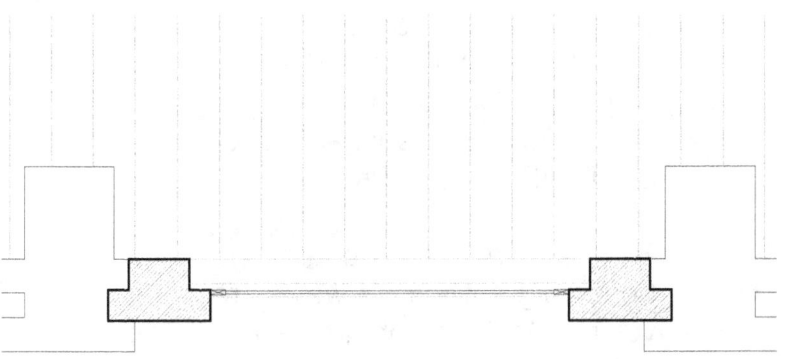

0   0.5   1   2

BUILDING: Sint-Barbaracollege
ARCHITECT: Stéphane Beel
REALISATION: 2009
LOCATION: Savaanstraat 33, Gent
STUDENT: Alessio Profeta

The Savaanstraat is characterized by a number of big entities, as well as residencies. This volume, that fits right in like a sugar cube, has but a few big windows to open up the site to the street. Those windows have become like a landmark of the street. They immediately catch everyone's eye.
This window in particular houses a foyer (cultural room). From the outside, it leaves a longing to peek in, just to know what happens there. From the inside it leaves a view over the well-known street.

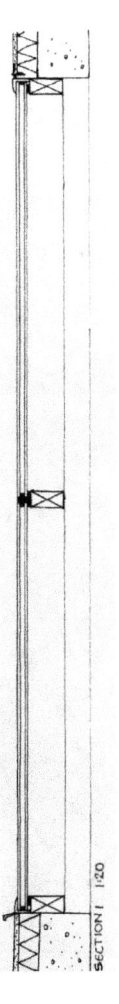

ELEVATION OUTSIDE 1:20

SECTION 1 1:20

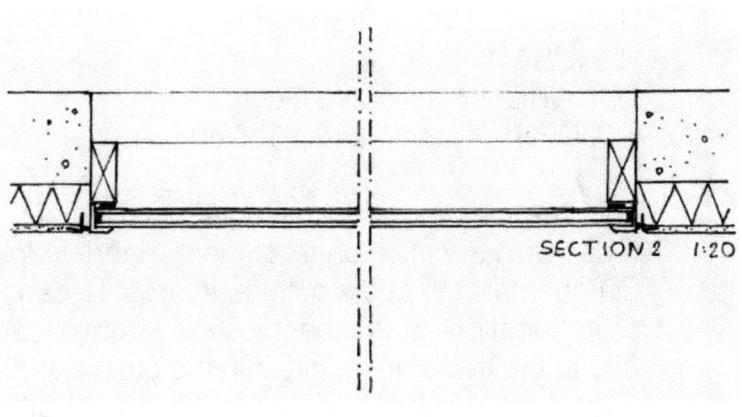

SECTION 2 1:20

BUILDING: Café het Spijker
REALISATION: 1734
LOCATION: Pensmarkt Gent
STUDENT: Antonio Estan Torres

This is a building which is located in one of the facades in the main canal of Gent, Handelsdok. On one hand, it has a cultural interest because of its location, in the past it took an important role in the trade of the city. On the other hand, I chose this window because of its structural interest and curious shape from the outside. It is built in a structural massive wall of bricks, where remains the stones who give the character to the façade composition. The main structure is made up of wood so the structural result is an interesting subject to study.

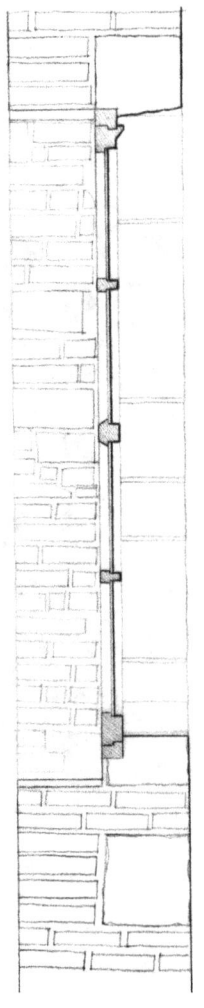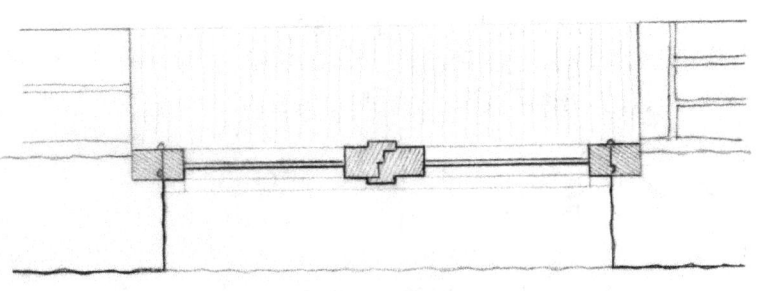

BUILDING: Private House
LOCATION: Predikherenlei 8, Gent
STUDENT: Cristina Janés Lloret

This house date back to the 16th-17th centuries. It has a stepped gable in traditional brick and sandstone architecture of the 17th century. In 1977 the house was renovated. In the same year the house was declared a heritage building.
It's usual to see in the city this kind of windows, with the composition of four windows and divided with a cross. In the 17th century they used to do a colourful frame windows like this one. This choice is because it's good to know how it works a typical window of the city of Gent.

BUILDING: Frank Steyaert's Atelier
REALISATION: 15th Century
LOCATION: 16 Tinnenpotstraat Gent
STUDENT: Dimitrios Antoniou

This triangular window was found at Tinnenpotstraat 16 in the historical city center of Gent. The window is positioned and shaped in such a way so that it becomes the most fundamental architectural element of this 15th century building that now belongs to the artist Frank Steyaert. The window becomes a symbol for the artist as well as a symbol for the building, giving it a unique identity.

BUILDING: Fallen Angles Store
LOCATION: Jan Breydelstraat Gent
STUDENT: Dimitrios Triantafyllou

This toy store is located in the commercial heart of the city of Gent. It's facade is divided in two identical and symmetrical windows. The interesting point is that those two windows contain two different kind of toys. In the first one, we can find vintage toys, all in brown-dark colours. In the second window, someone can see the modern toys, which are brighter and colourful. Those similarities and diversities, make those windows unique.

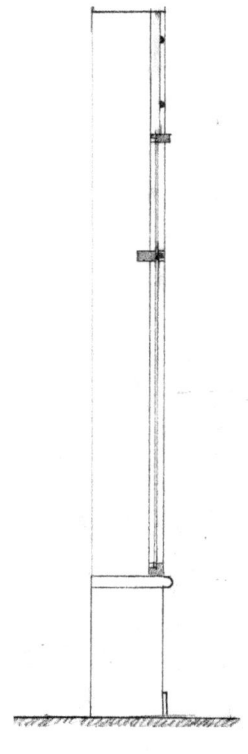

BUILDING: Desmet-Guequier cotton mill
REALISATION: 1830
LOCATION: minnemeers 9 Gent
STUDENT: Elisa Monge Moreno

This building that is currently The MIAT Museum of Industry and Textile, was one of the most important industrial buildings in Gent. It was a significant cotton mill when the city was on the top of financial issue, and people from everywhere moved to Gent to get a job there. As it is a place for industrial activities, the whole building is made into another scale. The size of the ceiling, structure and windows had to be in a bigger scale than it could be the family scale of a house, it had to be the scale of community. That's why the materials had to supply the heights and the presence of this building.

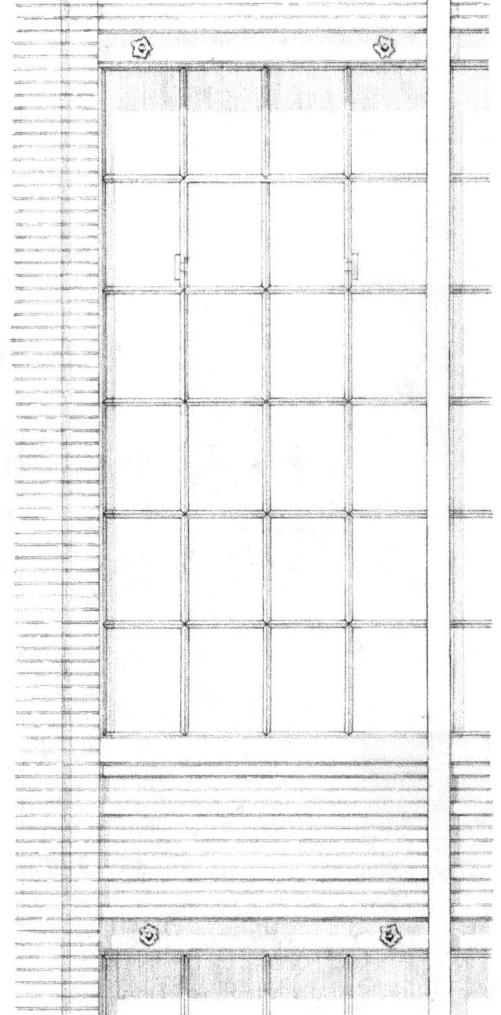
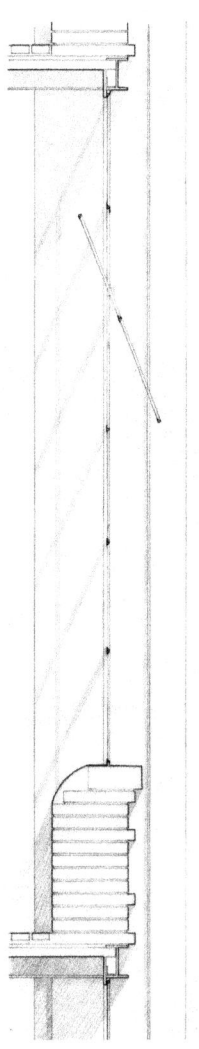

BUILDING: Poëziecentrum
REALISATION: 1852
LOCATION: Vrijdagmarkt 36,Gent
STUDENT: Inigo Uribarri Parada

This window at Korenmarkt has witnessed 800 years of history of Gent. Thanks to its central situation in the city has witnessed markets, festivals, exhibitions, and more importantly has seen the city evolve during its history.
Built in traditional stone wall consists of about five feet thick, which reflects a massive image in their structure fachada.The arch above the window distributes the weight in the front, and th e pillar located in the center of the window also acts as etructural compositional element. The building its a mix used building with a brasserie and some dwelling on it. Because of the modern regulations, it has been refurbished, aesthetically and also technically.

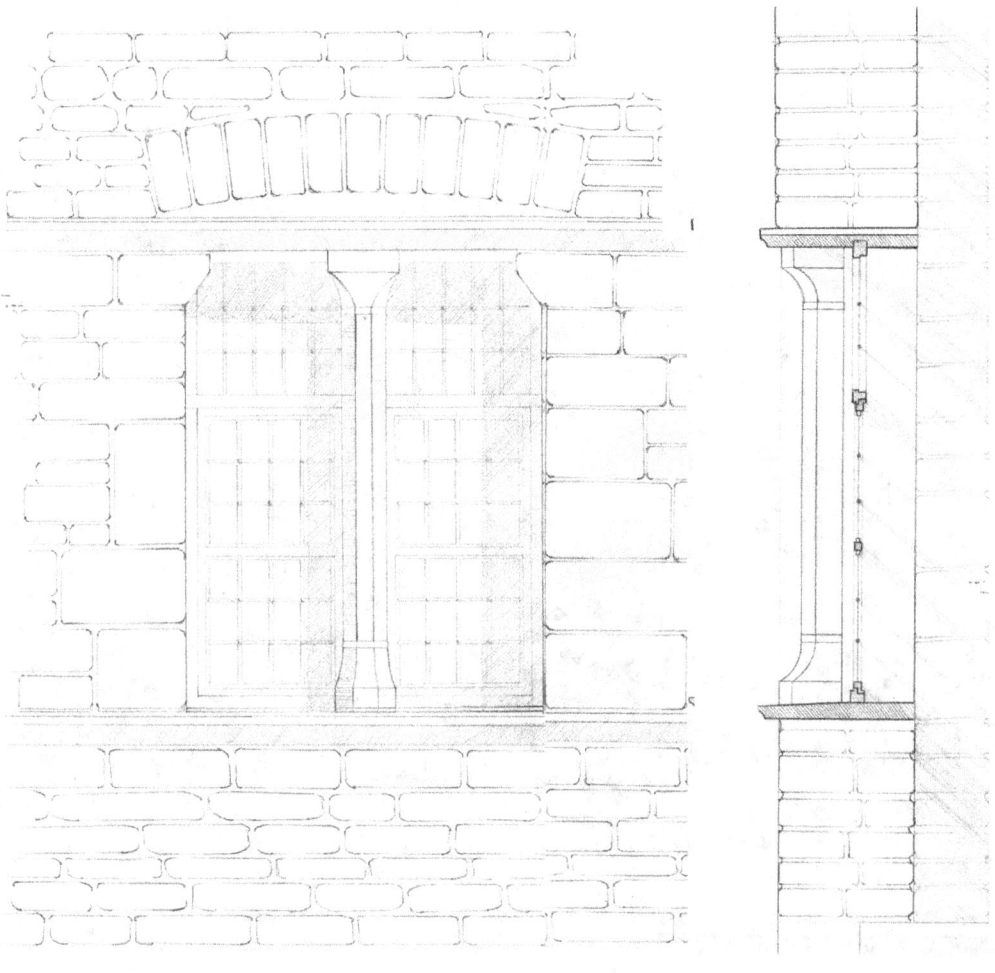

BUILDING: Nucleo Kunstenaars atelier
LOCATION: Lindelei Gent
STUDENT: Laura Gorina

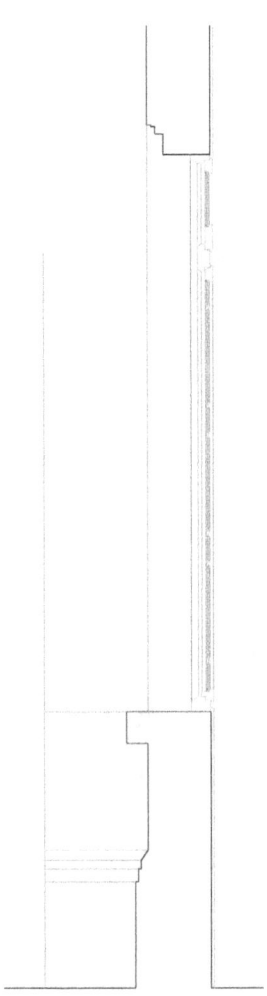

BUILDING: Coffee Max
REALISATION: 1715
LOCATION: Goudenleeuw plein 3 Gent
STUDENT: Martin Kips

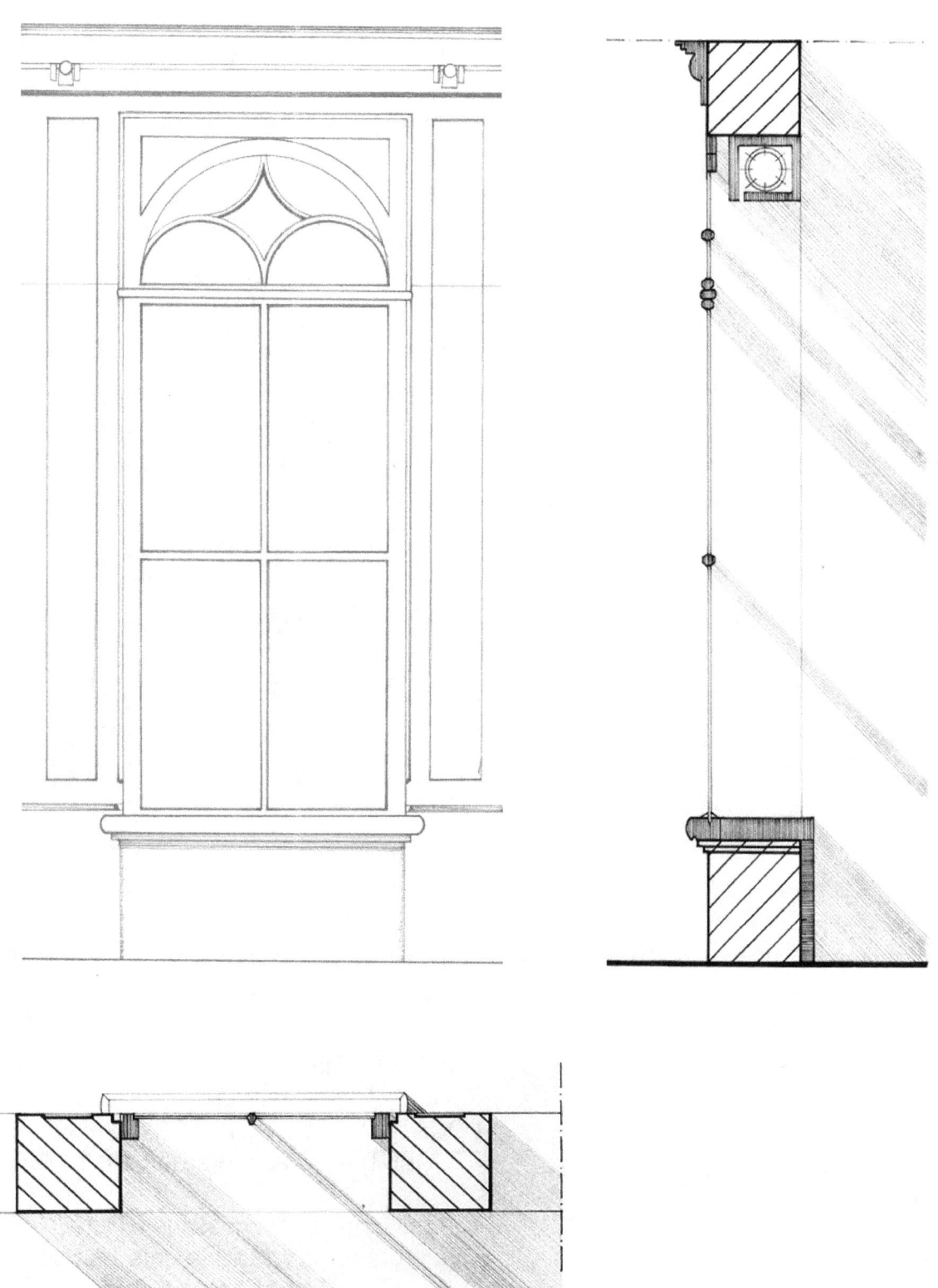

BUILDING: Platte Beurs coffee
REALISATION: 16th century
LOCATION: Klein Turkije 20, Gent
STUDENT: Maximo Gerardo Santos Martinez

The building is known as the "Platte Beurs" because of the name of the café that used to be in the place. During the 1900s it went through a complete restoration work which gave as a result the façade we can still see today. The "t" form-window composing the building is typical from the 176th and 17th century. Therefore, the choice of this window and the analysis through plans, sections and history will allow a better understanding of this element that we can see almost all over the centre of the city of Gent.

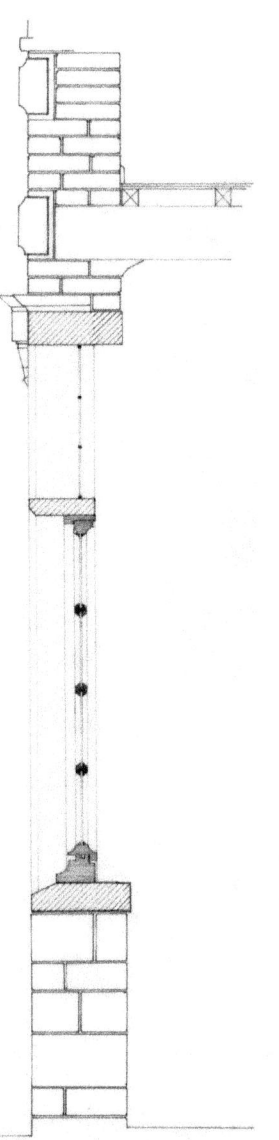

Group 06

"Our House" Socialist Worker
Groot gerechtsgebouw
Abby from Sint-Lucas
UFO gent
Les Ballets C de B
Private House
Sint Pieter Station
Proud of House
Old courthouse
Pakhuis Clemmen

BUILDING: "Our House" Socialist Worker association
STUDENT: Arnout De Schryver

Gent was in the second half of the 19th Century an example for the modern workman unions. Soon the socialistic union builded two "temples" for the workmans and their unions in the city. One at the "vooruit" the other at the "vrijdagsmarkt". Both buildings where build by the architect: Ferdinand Dierkens. Money was not an issue in his architecture. The workman only deserved the best materials, the most beautifull looking building, etc.. The building was constructed with the best and most innovative materials. The window in front of the building gave a bit the feeling of a trainstation (in that time, the symbol of the fast growing, developping society).This window is more than a window in it's smallest meaning.* This window was construced to create a status to a complete building. This window, toghether with the function of the building, explains in one frame what was happening in the beginning of the 20th century.

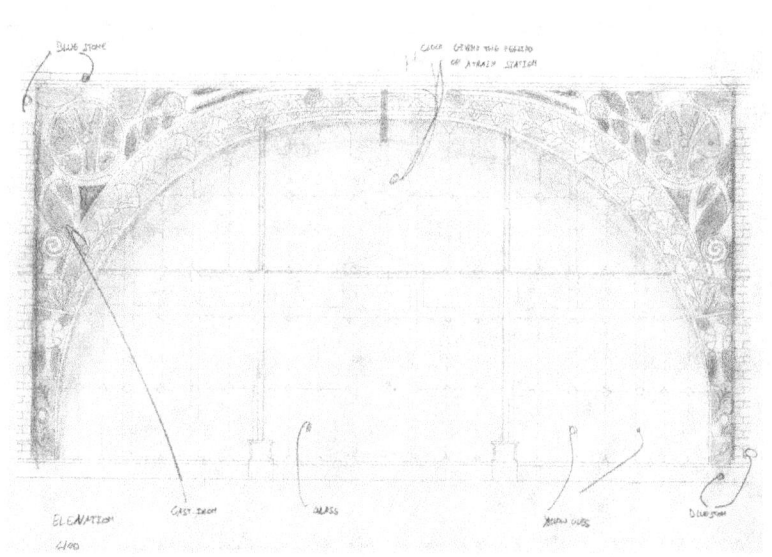
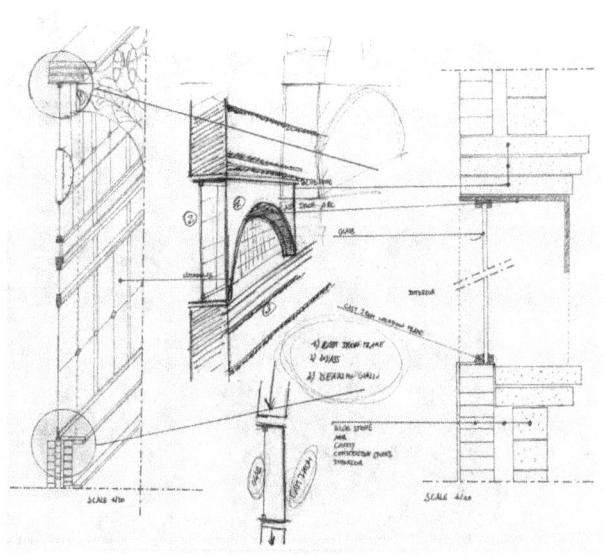
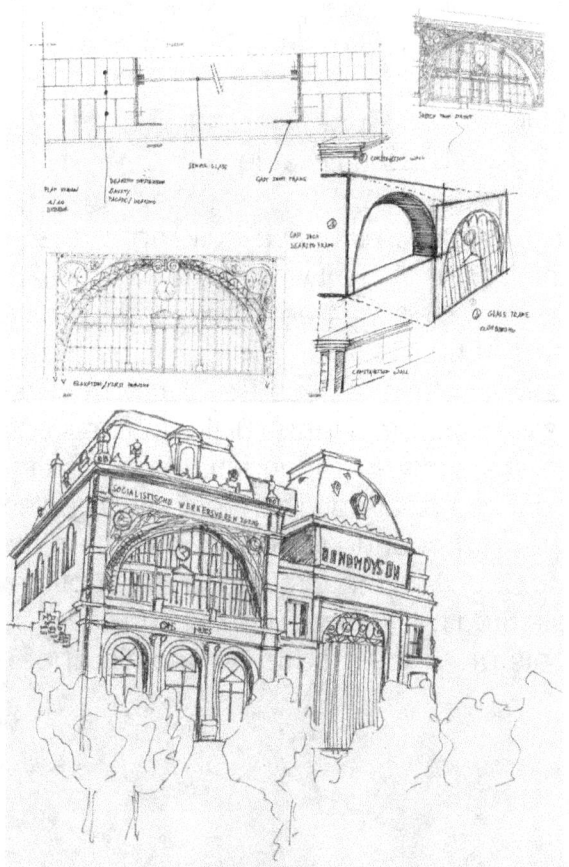

BUILDING: Groot Gerechtsgebouw
ARCHITECT: Stéphane Beel and Lieven Achtergaal
LOCATION: Opgeëstenlaan Gent
STUDENT: Ben Delabie

This large courthouse building in the north of the centre of Gent was just finished in 2007 and is designed by renowned archichitects Beel and Achtergaal. On this site used to operate a railway station. Because of the decline of the textile industry in the 70' the transport of goods to this site was redundent. As a result the station was aborted. The most striking element of the new courthouse building is the glass facade. It shows the circulation the public over the stacked pathways at the outside of the building. This gives the facade a dynamic appearance.

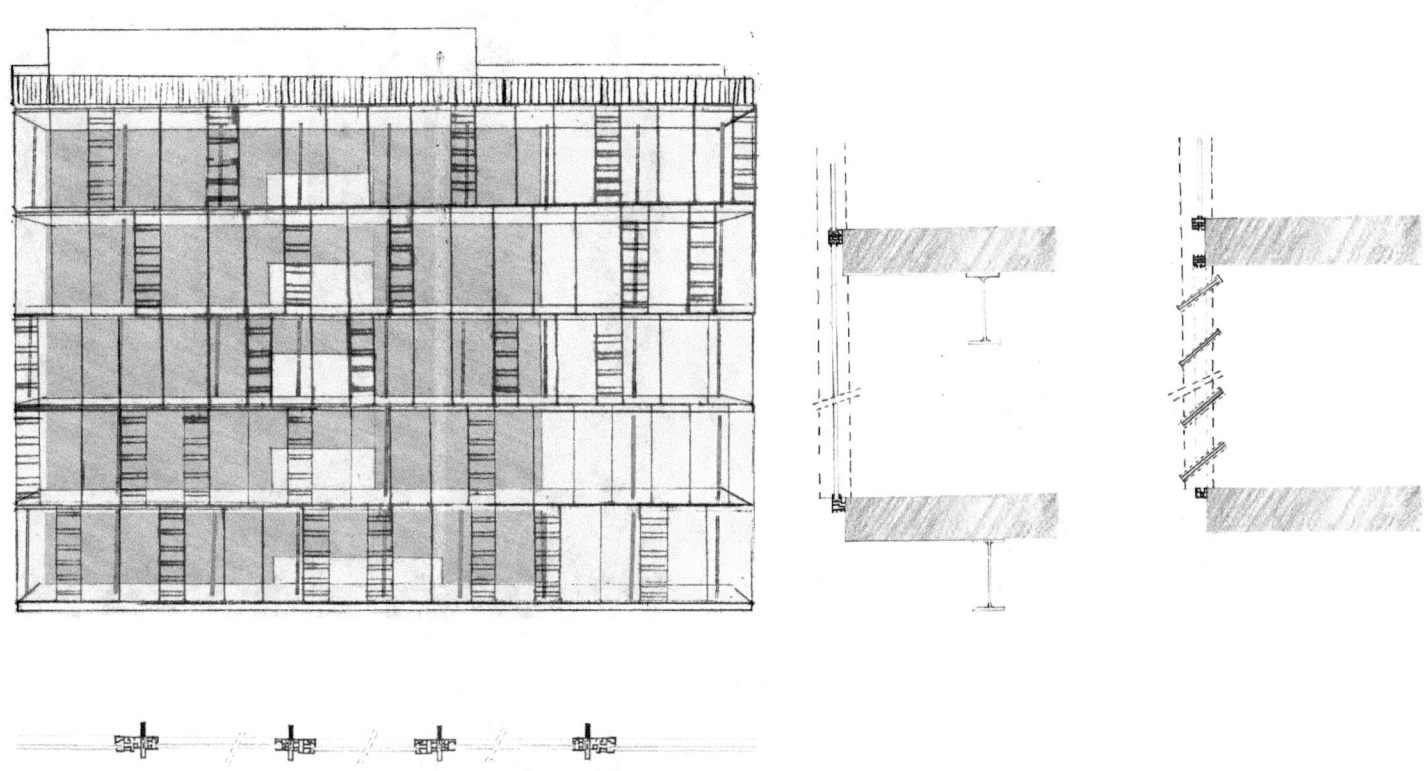

BUILDING: Abby from Sint-Lucas
REALISATION: 1146
LOCATION: Hoogstraat 51, Gent
STUDENT: Caroline Melders

From abbey to architecture school
In the 12th century, there was a lot of leprosarium so the city of Gent bought a plot of land from Sint-Baafs abbey, located in Marialand, to build a residence for people with leprosarium. Leprosarium disappeared in the 16th century, that's when the city of Gent decided to start a charity school in the empty parts of the building. From 1794 it went downhill because the abbey had to pay really high taxes. The nuns who were living there had to leave and in 1809 there was a definitive departure of the nuns. After that period, the abbey had a lot of different functions such as a cotton mill, charity workhouse and eventually it became Sint-Lucas, academy for science and art in 1995. I have chosen this window because it intrigues me. It looks so beautiful with the detailing in the window and it's interesting when you look through the window to see the inner garden. It gets even more interesting when you read the history behind this building and its surroundings.

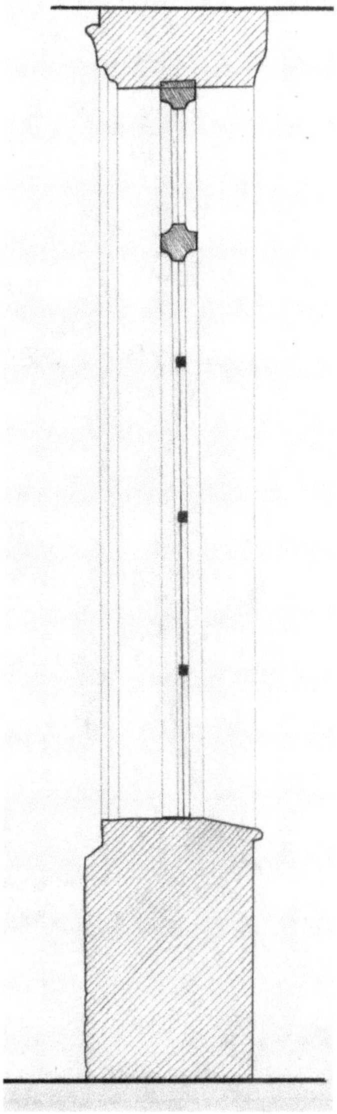

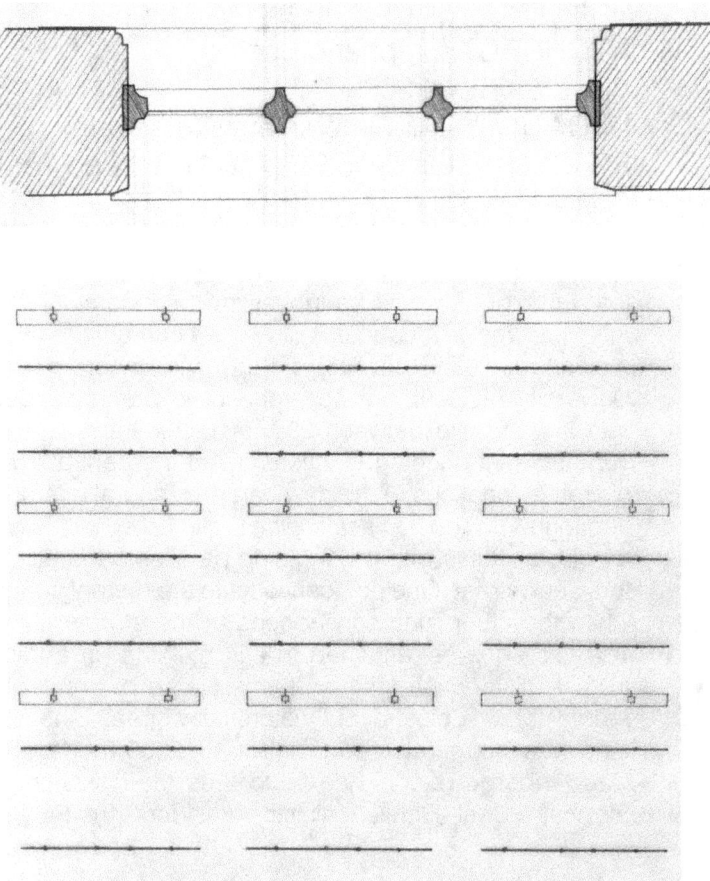

BUILDING: het UFO uGent
ARCHITECT: Stéphane Beel
REALISATION: 2009
LOCATION: Sint-Pietersnieuwstraat Gent
STUDENT: Francisco Pedro de Carvalho Sant Costa

The reason why I have chosen this building's window is that I thought it would be interesting to understand how this new type of facade is build so we can achieve such a clean result.This building is the University forum of the University of Gent, whose architect is Stéphane Beel, and it was build between 2007 and 2009.It host a several student-oriented functions, all of the central administrative functions of the new campus, and also a great auditorium, with 1000 seats, which can be split in two by a cell wall.It lies with its long side parallel to the St. Pietersnieuwstraat. One end connects to the rectory square, while the other (one floor) opens at the student plaza. The two squares, attached to this building, in addition to their distribution function forms a kind of urban vent rooms for the closed St. Pietersnieuwstraat. The building was inaugurated on October 15, 2009 by Paul Van Cauwenberge rector, vice rector Luc Moens, architect Stéphane Beel, Flemish minister-president Kris Peeters.

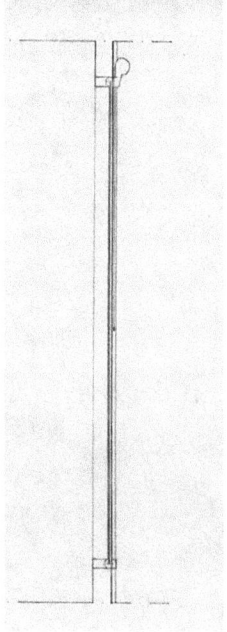

BUILDING: Les Ballets C de B
ARCHITECT: Architecten De Vylder Vinck Taillieu
REALISATION: 2008
LOCATION: Bijlokekaai Gent
STUDENT: Jim De Pan

Because the facade of the building is made completely out of glass (curtain wall), the building seems really light and transparent. At the meantime, the façade reveals everything behind it: construction, program, internal walls, openings, ... The border between public and private is made really thin. Especially the aspect of placing the red brick walls, right behind the glass facade, triggers me. It's an unusual view but for sure not less interesting.
 I've chosen a part of the facade that concludes an brick wall right behind the curtain wall, een open space, concrete and steel structural elements and an opening window.

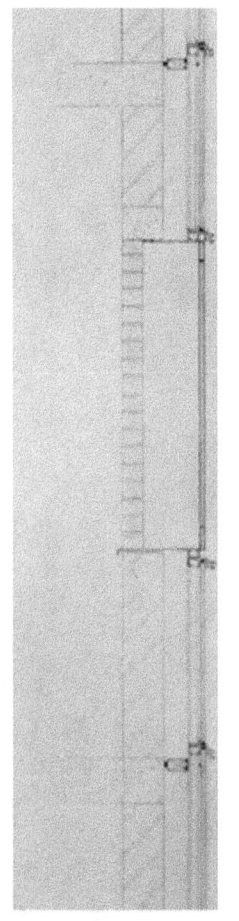
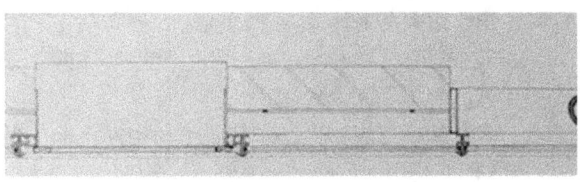

BUILDING: Private House
ARCHITECT: Félicien Bilsen
REALISATION: 1925
LOCATION: Prinses Clementinalaan Gent
STUDENT: Kasper Denayer

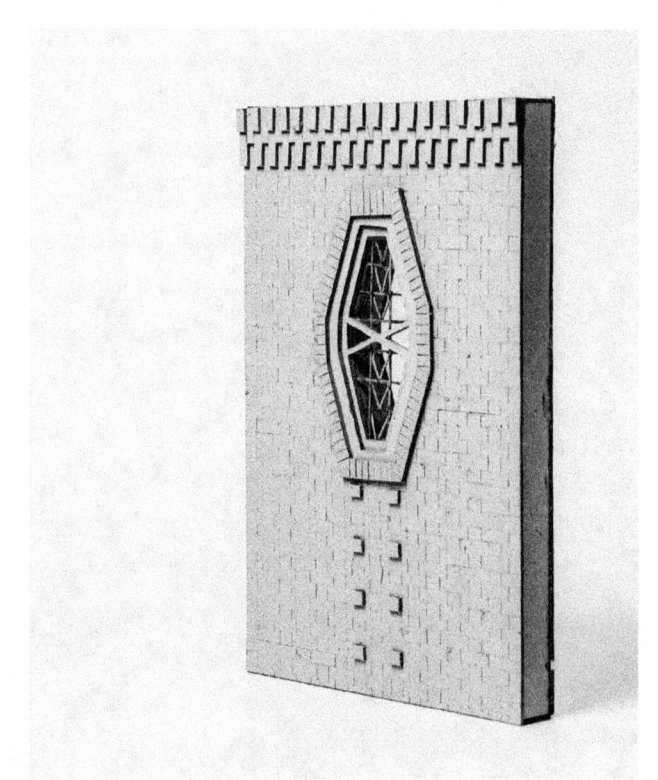

BUILDING: Sint-Pieters Station
ARCHITECT: Louis Cloquet
REALISATION: 1908-1913
LOCATION: Gent
STUDENT: Keshia Mannaert

The first so-called "Small Sint-Pieters", a stop on the Brussels-Ostend, was introduced in 1889-90 and was at the height of the current Park Square. A new station was in the develop¬ment plans of the southern districts early 20th century provide the Sint-Pieters-Aaigem. The realization of the new station in the immediate district was also accelerated. Engineer architect Louis Cloquet was circa 1908 commissioned to design the new sta¬tion. The final design was carried out in eclectic style, characteristic of L. Cloquet and inspired by medieval national elements. The rich material was determined at national level. I have chosen this building because of all the detailed elements you can find in the façade. For example there are 5 different types of stone used, which gives this building a particular characteristic. As through the years I've "used" this building a lot and each day I'm able to find new elements that I didn't re-cognized before. It's like a road trip through the buil-ding itself.

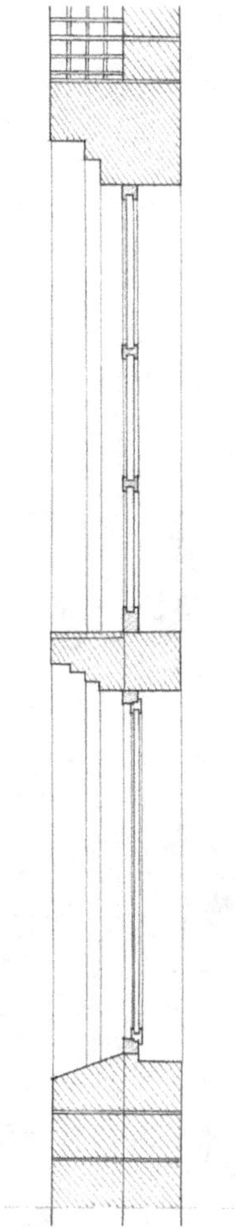

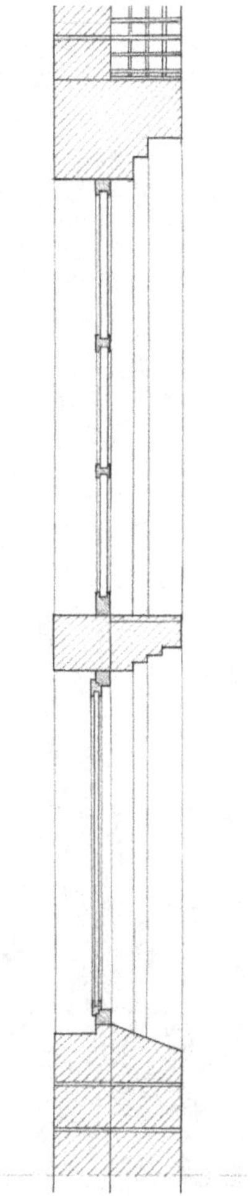

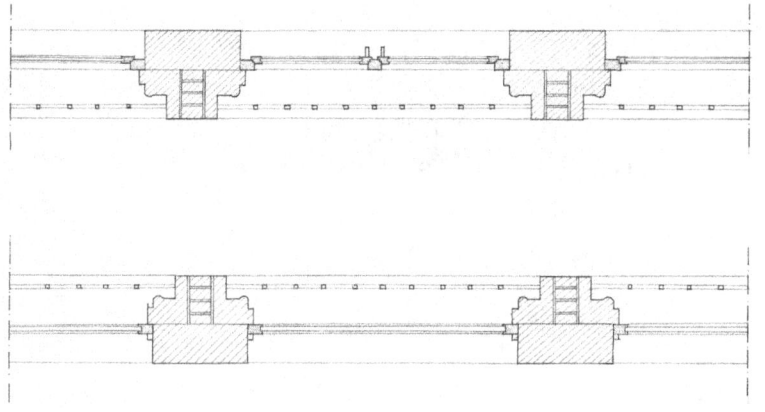

BUILDING: Proud of House
ARCHITECT: OYO
REALISATION: 2012
LOCATION: Merelbekestraat Gentbrugge
STUDENT: Michaela Pivonkova

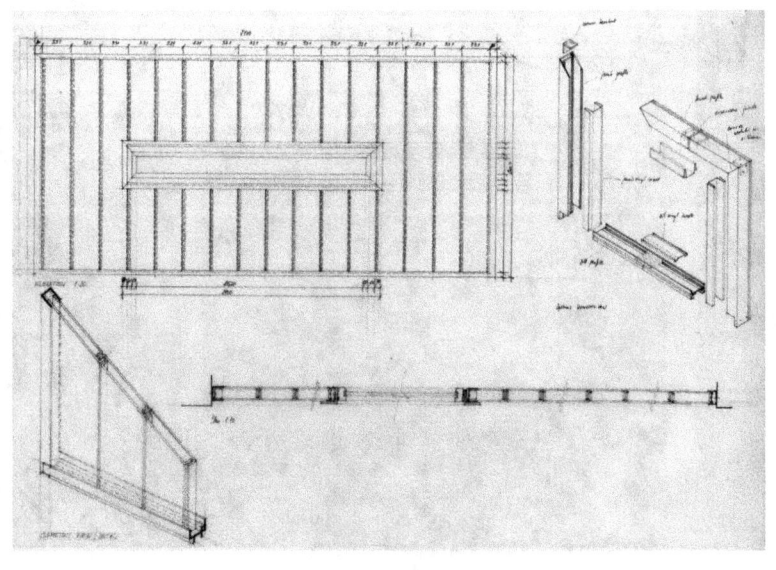
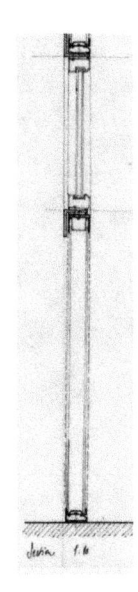
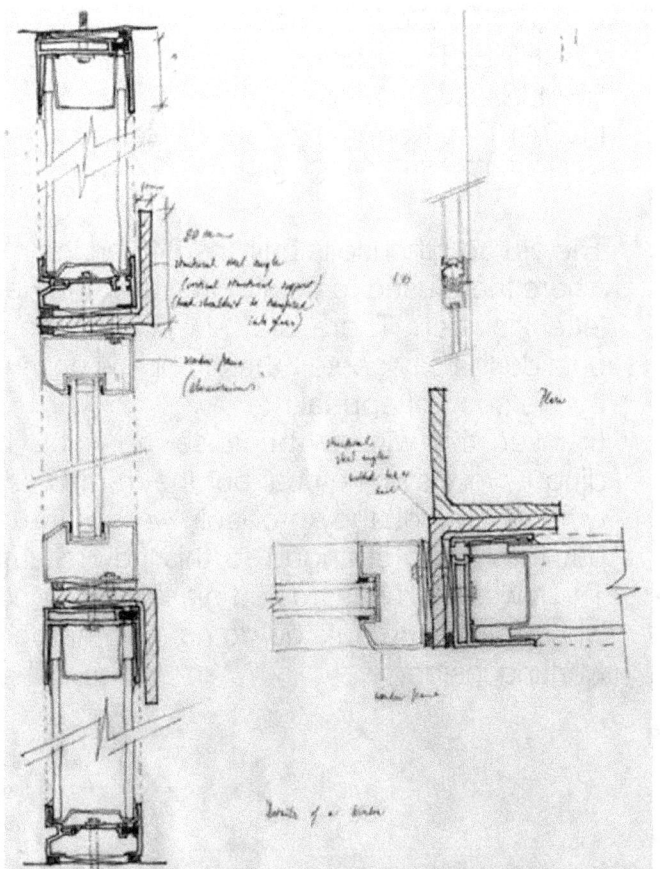

BUILDING: Old Courthouse
ARCHITECT: Lodewijk Roelandt
REALISATION: 1846
LOCATION: Koophandelsplein Gent
STUDENT: Nina De Coster

The old courthouse is built in 1846 on the place where there used to be a monastery, the Recollettenklooster . There is a new courthouse, just outside the city center, the old one is only used by the court of appeal.
I picked this window because the whole building is very symmetrical but the frames of this window are not, they probably were in the past but they were changed to this new situation. The lowest part has frosted glass so the people who are walking outside do not see the people working inside.

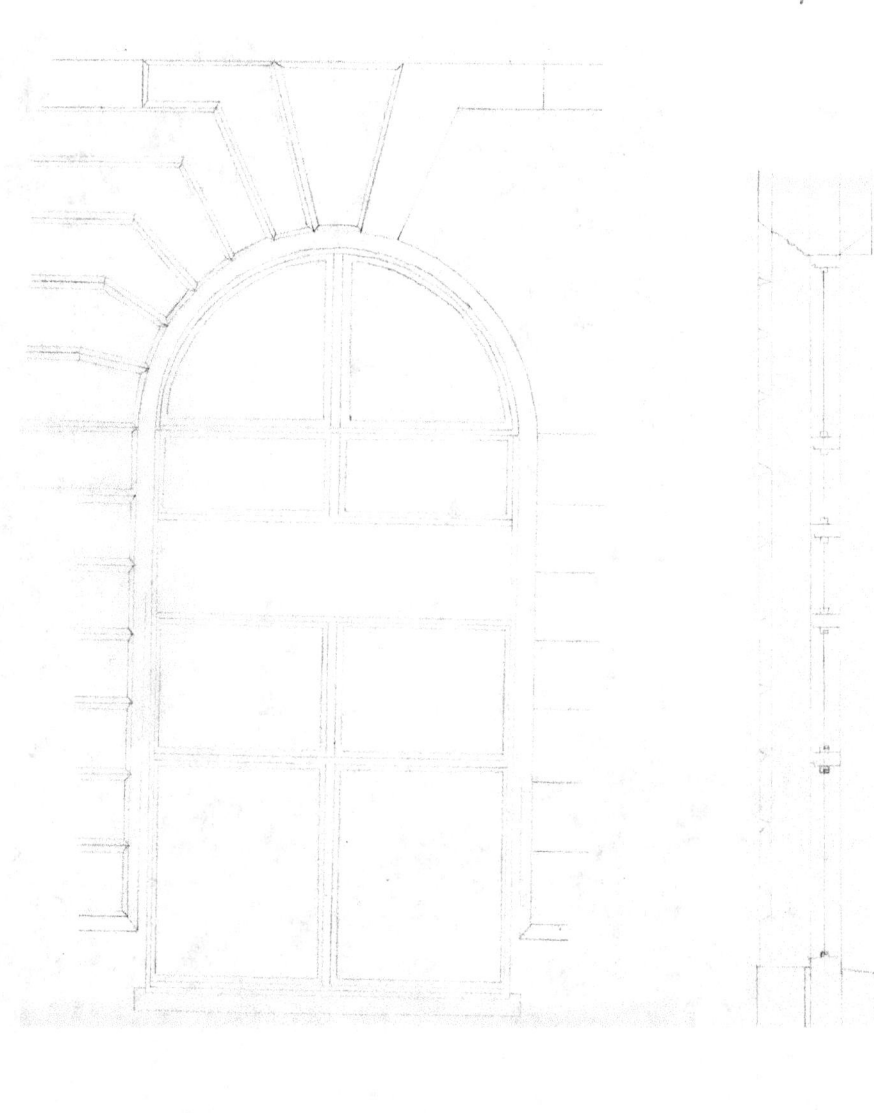
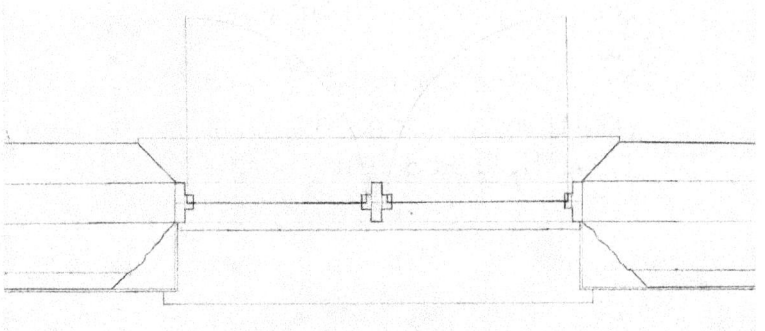

BUILDING: Pakhuis Clemmen
ARCHITECT: ALT architecture office
REALISATION: 2012
LOCATION: Veldstraat 82a Gent
STUDENT: Olga Salata

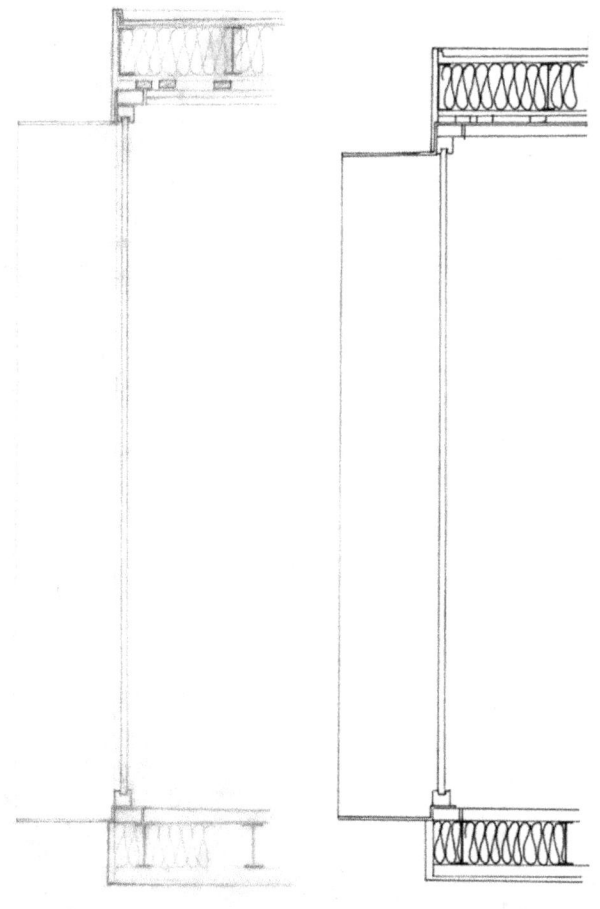
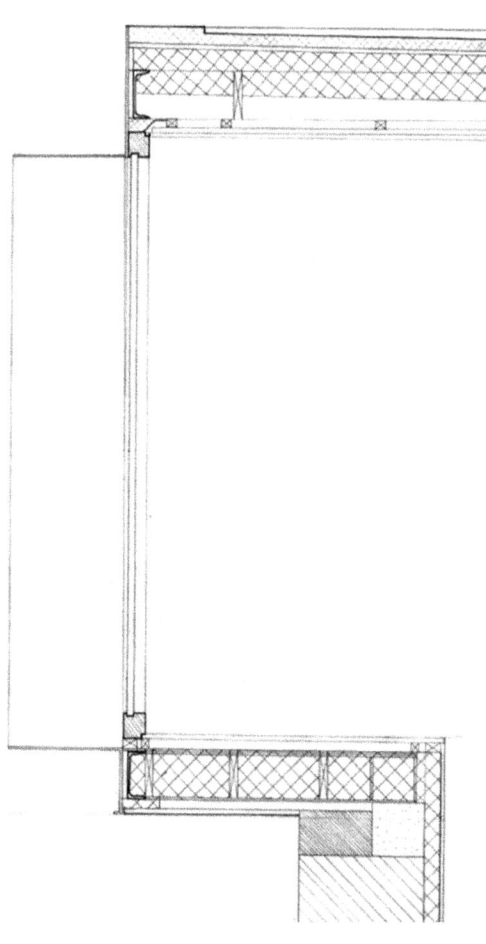

125

Thanks to all the participants of the workshop.

Students

Barbara Polakova
Borkowicz Iwo
Emiel Furniere
Gonçalo Oliveira
Helder F.Ferreira
Igor Machata
Maja Berden
Ban Ngoc Tran
Dalia Brandone Gonçalo Jakub Senkowski
Jolien Vandereecken
Jonas Werkbrouck
Marika Piekarczyk
Martin Mikovčák
Petra Ross
Radka Vilímková
Stepan Vasut
Tomas Matata
Anouk Desmedt
Dang Phuong Dan Tran
Erika Vandekerckhove
Ghaith Khallouf
Kaitaro HiraiMarika
Khanh An Thi-Phan
Minh Ha Nguyen
Danh Nguyen Thi-Kieu
Takeshi Uchida
Yuki Fujita

Alessandro Martinelli
Ilona Pietrus
Kalin Hristov
Kateryna Taratynska
Mandana Fouladi
Zhana Ivanova
Alejandro Domínguez
Alessio Profeta
Antonio Estañ
Cristina Janes
Dimitrios Antoniou
Dimitrios Triantafyllou
Elisa Monge
Iñigo Uribarri
Laura Gorina
Martin Kips
Maximo Santos
Arnout De Schryver
Ben Delabie
Caroline Melders
Francisco Costa
Jonas Werkbrouck
Jim Du Pan
Kasper Denayer
Keshia Mannaert
Michaela Pivonkova
Nina De Coster
Olga Sałata

Tutors

David Kohn
Dieter de Vos
Bram Aerts
Patrick Moyersoen
Bart Van Gassen
Steven Geeraert
Johan Nielsen

www.ingramcontent.com/pod-product-compliance
Lightning Source LLC
Chambersburg PA
CBHW081047170526
45158CB00006B/1890